The Science of Consciousness

Eva Déli

The Science of Consciousness

How a new understanding of space and time infers
the evolution of the mind

2015

The Science of Consciousness

How a new understanding of space and time infers
the evolution of the mind

For information, address:
Eva Déli
[Nyiregyhaza 4400, Hungary, Benczur ter 9
Library of Congress Cataloging-in-Publication Data

Déli, Eva
The Science of Consciousness
How a new understanding of space and time infers
the evolution of the mind

Includes Bibliographical references
and index

ISBN 978-963-12-2626-3

Printed in U.S.A./Hungary

5 4 3 2 1

„ I dedicate my book to theoretical physicists, who came to understand a tangible world only through mathematics and believed in its existence unseen, without experimental evidence of any kind. That is true faith, the noblest human drive for the truth. "

Table of Contents

List of Figures

Preface

„ Along with this disregard for historical linkage there is a tendency to forget that all science is bound up with human culture in general, and that scientific findings, even those which at the moment appear the most advanced and esoteric and difficult to grasp, are meaningless outside their cultural context. A theoretical science, unaware that those of its constructs considered relevant and momentous are destined eventually to be framed in concepts and words that have a grip on the educated community and become part and parcel of the general world picture—a theoretical science, I say, where this is forgotten, and where the initiated continue musing to each other in terms that are, at best, understood by a small group of close fellow travelers, will necessarily be cut off from the rest of cultural mankind; in the long run it is bound to atrophy and ossify, however virulently esoteric chat may continue within its joyfully isolated groups of experts."

—Erwin Schrödinger

In relationships, incest is rightly abhorred. However, a tendency exists for a sort of intellectual inbreeding in relationships, in which people only associate with those with who they are already in agreement. Such intellectual isolation in the sciences is particularly dangerous because it limits our mental world to a relatively small circle of fellow scientists, and beliefs can easily be treated as „truths." Professional relationships initially stimulate thinking. Scientists even develop their own language; to outsiders, it sounds like a foreign language and, because it excludes some, it tends to segregate people. A field walled in by language turns inward. Gradually it becomes static and unfriendly to fresh ideas. By the time a young mind acquires the language of a field; its thinking is narrowed to within the prescribed mental boundaries. This limited view becomes more poignant by looking back in time to „scientific ideas" of a hundred—or even twenty—years ago. For example, Wegener plate tectonics and Mendel's ideas of genetics were shunned by scientists at the time.

Fritz Zwicky calculated the existence of extra matter (dark matter) in 1933, but his findings were not taken seriously until sixty years later, when dark matter became fashionable science. Established careers are anchored in accepted beliefs, which is antagonistic to true change. The peer-review process, which is intended to produce excellence in scientific literature, has been shown numerous times to fail when confronted with the best and most pioneering work (Peplow, 2014).

The three chapters that comprise this book cover a wide range of topics from theoretical physics to neurology to evolution. This book endeavors to show the connectedness between these separate and seemingly disparate fields from a single perspective. The greatest realization is the immense and deep operational and structural similarity and relatedness within the material and human world, and in the physical and social sciences. This work proposes that the mind, the universe, and even elementary fermions all share identical operations that stem from analogous energy structures, and that the visible, measurable world is only a tiny part of existence.

In this book, I present the current scientific understandings but intentionally leave out confusing, overly complex subjects. I introduce new concepts only where absolutely necessary. Hopefully, this approach will allow the pieces to fall into place for the reader. It is possible to conduct thought experiments even about string theory. Likewise, the ideas analyzed here are imaginative mental ponderings that attempt to answer some of the most pressing questions of theoretical physics, cosmology, neurology, and evolution. This work attempts to stitch together these vastly divergent fields as if they were part of a giant puzzle. Nevertheless, due to space constraints and my own ignorance, many excellent publications have been left out, for which I apologize. The credit for this book therefore belongs to the scientists who, by their mathematical prowess, created the basis of these ideas. The mathematics is difficult, abstract, and counterintuitive: only the individual genius of theoretical physicists could have brought forth such detailed understanding of elementary particles, often prior to experimental verification. Other phenomena cannot be measured at all, or cannot be measured in a mechanically feasible way. Although this book is merely a synthesis of their understanding, I do propose some ideas that go against the current scientific understanding, but not against experimental data. In fact, scientific data and common experience easily verify the main points of this book. These concepts are presented with the help of simple drawings; no equations are required. Every process is investigated in light of the change in energy and entropy, based on the principle of least action. This takes everything back to the roots of physics: the study of matter and energy. In my view, this approach is most prudent, because everything that occurs in the world does so only as a result of energy exchange. Traditionally, physics is a science that can be easily visualized because its concepts are connected to real-world processes. However, the introduction in the twentieth century of quantum mechanics changed things. Physics became more abstract and its reliance on difficult mathematics made it less intuitive and almost impossible to visualize. The introduction of string theory only made things worse; equations became too difficult to solve, approximations had to be used, and any „solution" to these approximate, fudged equations could never be tested because the strings were too small to measure. Like a train running without tracks, science entered shaky ground. A fractured science, physics has arrived at an impasse. A kaleidoscope of bizarre ideas are contemplated, such as the existence of multiverses, wormholes, branes (indicating subspaces formed by „membranes" of various dimensions), realities that exist only as mathematical extrapolations or only in the imagination. But the truth is often more shocking and awe inspiring than fiction. It is natural to be curious about our surroundings, and during the twentieth century our neighborhood has expanded to include the whole universe.

The following ideas might conjure up a surprising and even astonishing (but organically connected) world without the mysteries that unexplained phenomena might allow. But they just might lead to an easy way to build a universe, should we ever want to. So sit back and relax: the following is a user's manual for a simple universe, populated by intelligent beings (people, if you like). In this book, I propose a fundamental shift in our conception of reality: one that offers an overarching and coherent hypothesis. Its main points are yet to be tested, but already it can spark new ideas for scientific progress and offer practical ideas for mental excellence. Today the vast majority of us follow the random trajectories of our haphazard existence, managing day-to-day emergencies as dictated by circumstance. Sages and philosophers have intuited correctly how to live a fulfilling life but, without any scientific proof, these possibilities could not be precisely explained and have remained out of reach for most people. This new hypothesis gives a theoretical background and proposes a practical path toward happiness, purposeful living, and social success.

The ancient Greeks believed that the movement of the celestial bodies formed harmonies; the music of the spheres. They were ahead of their time. When the universe is examined on different scales, at different angles, or from different aspects, the same underlying regularity and harmony can be found: the music of the universe. This is a most fantastic realization.

Acknowledgments

Describing a new hypothesis is a difficult, audacious job. Vocabulary has to be refurbished with new meaning or created anew. Experience is lacking and imagination can easily lead you astray. Expressing ideas about a new scientific frontier requires audacity and involves a risk of failure. My children, Zoltan, Veronika, and Gregory have been my greatest inspiration during the preparation of this work. There were so many people who aided me in various ways on this sometimes arduous path. My mother graciously let me stay with her during the final preparation of the work, removing any existential worries. I am thankful for my editors Brett Kraabel and Julie HawkOwl, who worked diligently and with care on the manuscript. Péter Futó is a trusted mentor, whose encouragement, counsel and support was instrumental in clarifying my goal with increasing focus. Károly Bonyhádi designed the elegant book cover and the illustrations were done by Tamás Krajecz at Lionardo. Valuable consultations with Bertalan Kercsó were essential for better insights into theoretical questions. I also thank Doug Myers and Dénes Dudits for their encouragement and support.

How to Read This Book

"What is essential is invisible to the eye."
—Antoine de Saint-Exupery

Forget your picture of the universe. Our sight, hearing, indeed all our sensory experience and even measuring equipment have been leading us astray.

I recommend this book to graduate students and professionals in physics, biology, neurology, medicine, psychology, sociology, economy, engineering, and related fields. I also recommend it to anyone with an open mind who is interested in the cosmos and our place in it, and to those who are curious about evolution and the mind. It is also helpful for anyone searching to understand the deep forces governing human relationships. I recommend it for parents raising children, leaders working with employees, and all those with ambition and hunger for success. The ideas presented herein represent a radically new physical world view which, if proven correct, will change our fundamental understanding of the world and the mind. The book contains three chapters; each built on the preceding one. Reading the chapters in sequential order is therefore strongly recommended. The chapters are structured as collections of scientific papers that contain an abstract, discussion, and conclusion. I also included summary after each subsection. This organization should facilitate an overview of the important points and thereby aid comprehension.

Admittedly, this book is not an easy read. It will challenge your mind by pulling the secure, twentieth-century world view out from under your feet. It replaces a static world view with a dynamic, constantly changing horizon that never lets you fall asleep. Bold-faced key terms, such as the Calabi–Yau torus , are defined in the Glossary at the end of the book. Extra information that might aid in visualization or understanding, but not strictly part of the discussion, is put in text boxes.

To describe specific subject matter, the book often uses language unique to the given scientific field. These words can often be glossed over without significantly impairing understanding. Just as you can use a television set or refrigerator without knowing their technical background, you do not need mathematical or scientific training in these concepts to grasp their practical application. Undeniably, I lack a complete understanding of all subjects discussed in these pages. Considering the wide range of fields talked about, such is the best we can do, and it reinforces my belief that collaboration in the sciences is imperative.

The aim of this book is to uncover the deep, organic connection between seemingly disparate fields. I have formulated a new hypothesis, a new coherent physical world view that has the potential to open new prospects in cosmology, particle physics, evolution, and the understanding of the mind. This new understanding will allow us to predict physical phenomena and emotional behavior. By introducing a mental map, success becomes possible in almost any field. The hypothesis also widens the implication of the evolutionary process and introduces societal evolution. Societal evolution is a deterministic process that converges toward democracy, peace, and the well-being of all humankind.

ONE

A NEW UNDERSTANDING OF SPACE AND TIME

Abstract

Over the past century, enormous progress has occurred in theoretical physics, following largely the path embarked upon by the theories of general relativity and quantum mechanics. In the last few decades, however, progress has slowed, diminishing hopes of reconciliation between general relativity and quantum mechanics. The following pages are a new attempt to overcome this impasse by examining entropic changes in a two-dimensional universe formed by **orthogonal** energies: space and time (macro-and microdimensions). Big Bang is an energy accumulation of chaotic string fluctuations, which formed an information-blocking horizon, dividing space into micro and macrodimensional subsystems. Within the microdimensions, called the **Calabi–Yau space**, standing waves form **temporal energy**. The closed microdimensions form minimal surface foam. Interaction with space changes the standing-wave frequencies (degrees of freedom), which is then recorded as information and called **time**. Aided by the **Pauli exclusion principle**, interactions gradually formed gravitational fields, increasing the differences in spatial curvature on a universal scale and forming the poles. The universe is bounded by its poles, the black holes and white holes. Due to their immense gravitational-field strength, black holes stabilize space, whereas the self-reproducing singularity of white holes decreases field curvature through spatial expansion. The minimal surface formation by the microdimensions and the increasing spatial curvature differences of macrodimensions lead to self-regulation. This idea eliminates the contradictions between general relativity, quantum mechanics, and string theory and leads to a deeper understanding of time, **mass**, **gravity**, and **entropy**. It introduces an intuitive structure of **black holes**, **white holes**, and the **galactic environment** and explains the accelerating expansion of the universe. Several contradictions that plague contemporary physics are explained within its framework, and the main points of the hypothesis can be verified by technically feasible experiments. The first section of this chapter describes the stringy mechanism of the Big Bang. The second section explains the proposed mechanism for the unification of string theory, general relativity, and quantum mechanics. The third section proposes a detailed mechanism for gravity and other elementary forces. The fourth section concludes, and notes about some quantum phenomena makes up the fifth section.

„Our present picture of physical reality, particularly in relation to the nature of time, is due for a grand shake-up–even greater, perhaps, than that which has already been provided by present-day relativity and quantum mechanics."

—Roger Penrose

Introduction

The Olympic swimming facility built for the 2008 Peking Olympics appears from the outside to be formed from giant bubbles. Its design was inspired by foam, which provides the most efficient way to partition space into cells of equal volume with the least surface area. On its largest scale, the cosmos shows a cellular structure that is unexpectedly similar to the elegant design of the Peking swimming facility (Alfvén, 1981). Various vacuum bubbles form a clearly defined, compartmentalized, foamy, sponge-like structure (Weygaert, 2007; Cantalupo et al., 2014) called the cosmic web. These endless formations of visible galaxies enclose us in a sphere with a radius of 13.8 billion light years, but that could be just a tiny sliver of an unexplainably uniform cosmos.

During the twentieth century, physics grew to rely on abstract mathematical modeling to uncover the deep structure of the material world and our cosmic origins. The result has been a spectacular success. However, the mathematics has become so difficult that special considerations are required, which often leaves them a merely formal significance (Dieks & Lubberdink, 2011). In addition, a hundred years on, inconvenient singularities have left general relativity, quantum theory, and string theory irreconcilable, leaving us with a number of inexplicable phenomena, such as the matter-to-antimatter ratio, the existence of extra matter (dark matter), extra energy (dark energy), and quantum entanglement. Gravity, a mysteriously weak force, still awaits an accepted theory and the presumed initial low entropy of the universe (i.e., the cosmos) cannot sufficiently explain the low entropy found today, the thermodynamic arrow, or the arrow of time. Although the behavior and qualities of space and time show vast and important differences, general relativity considers time a minor, fourth dimension to space. But unlike space, time is irreversible even within Euclidean regions and differs in accelerated versus nonaccelerated time frames. It is impossible to move in space without also changing temporal coordinates, whereas time continues even in the absence of motion. The marriage of space and time, officiated by Hermann Minkowski a hundred years ago, is ripe for a divorce.

Occasionally, science arrives at a junction where it is necessary to start from a clean slate. In what follows, I present a new hypothesis accompanied by some new vocabulary and substantiated with several examples. I consider temporal and spatial energies interlaced like horizontal and vertical threads on a cosmological loom, weaving the fabric of the universe (**Figure 1.1**)[1]. String theory (Veneziano, 1968), quantum mechanics, and the theory of general relativity are all manifestations of these orthogonal fields permeating all of space. Through their interactions, elementary forces emerge naturally.

Preliminaries

The Heuristic Foundation of Hypothesis

According to general relativity, clocks and time can only be properly defined in the presence of matter. In other words, interaction is an essential ingredient of time. Without interaction, time stands still. Here we take this even further and propose that interaction (decoherence) generates time. That time is a function of entanglement has gained serious support by the work of Moreva and her colleagues, who showed that change is the privilege of inside participants (2013). The mechanism of „static" time views entanglement as a clock system that allows one

---

[1] The field curvature cannot be expressed correctly in two dimensions. For this reason, its representation in the figures is not consistent and is for symbolic presentation only.

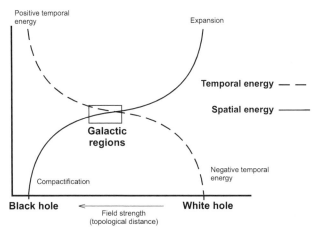

Figure 1.1. Relationship between spatial and temporal energies of universe   Black holes and white holes are the energy gate keepers of the cosmos. White holes are places of expansion, whereas black holes are formed by compactification. The equal balance of MaDV and MiDV energies shapes the galactic regions, where excess MiDV is experienced as gravity. Since no decoherence occurs at the poles, they can only be experienced indirectly.

to perceive the evolution of the rest of the universe. Deprived of such clock system, outside observers find the universe static and unchanging. Accordingly, interactions acting through entanglement would necessarily evolve toward polar singularities (Figure 1.1). The existence of poles, called white and black holes, also appears as the consequence of some solutions of Einstein's field equations. Black holes have been identified indirectly by observing their gravitational effects, but white holes remain hypothetical. However, the nature of white holes has been speculated as negative-curving, bulging fields, which deflect incoming energy, even light. Thus they are the source of illusionary light rays; a fact reflected in their name and which would pose a great challenge to their discovery (**Figure 1.2**.a and 1.1). In the following discussion the existence of both white and black holes is assumed.

String theory models particles as vibrating strings that arise out of microdimensions, called the Calabi-Yau space (Veneziano, 1968). Dividing a universe by a causal information-blocking horizon (manifold) into micro- and macrodimensional subsystems would permit only the formation of discrete frequencies. Entanglement acting through mirror symmetries and dualities of smooth horizon surfaces would always produce energy-conserving, symmetric spatial topology (Duncan et al., 2015). As a consequence, the state of the horizon or its energy level is determined by interaction. The close link between information and thermodynamics is expressed by the Landauer's principle, which recognizes that energy and information are convertible quantities. For the first time ever, converting information into free energy has

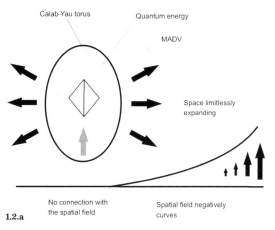

Calab-Yau torus

Quantum energy

MADV

Space limitlessly expanding

No connection with the spatial field

1.2.a

Spatial field negatively curves

Figure 1.2. Energy structure of white holes
The arrows indicate the direction of energy inside and outside the torus. The gray arrow shows the spin direction, the white arrows point out the effect of the contracting manifold inside the torus, and the black arrows indicate the forces outside the torus. (a) The Calabi–Yau torus at white holes. The information-free manifold is energetic and contracted. The quantum waves are nonexistent or weak and thus cannot separate from the energetic manifold. The compacted manifold forms MaDV, which insulates the torus, giving rise to negative spatial curvature and spatial expansion. (b) The structure of space at white holes. The contracted torus has no connection with the SF. The negative-curving field is weak and frayed. Expansion (Ricci flow) swells the fabric of space. Incoming quantum energy is channeled away.

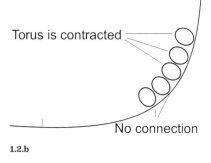

Torus is contracted

No connection

1.2.b

been demonstrated (Toyabe et al., 2010) and the exact amount of heat released when one bit of information was erased has been measured by Bérut and colleagues (2012). This dynamic variable, which can transform into information and back again through interaction is called quantum potential in Bohm mechanics (Bohm, 1952; Bell, 1987). However, **manifold energy** expresses the inherent non-local quality of the possible energy states of the system. In addition, standing waves can be considered **quantum energy** and their increasing frequencies are recorded as quantum information (information), thus manifold energy and quantum frequencies are inversely proportional. Quantum energy increase is recorded as information accumulation of the manifold and, inversely, decreasing quantum frequencies generate manifold energy, while releasing heat. This way, quantum energy is proportional to microdimensional volume, which forms the **temporal field** (TF), whereas manifold energy is proportional to the macrodimensional volume, which is space. The above idea successfully and irrevocably takes apart spacetime. Time is part of the microdimension, which leads to its fundamentally separate qualities and behavior. Information accumulation in the microdimensions occurs at a predictable pace, and turns time irreversible, whereas space can be navigated at will. However, energy-information exchange between space and time is the exclusive quality of interaction, when wave-function collapse forms singularities and shifts the quantum frequencies in sync with the field curvature. This way, energetic changes inside the torus, which nevertheless appears constant from the outside, are projected out to the field, changing its curvature and regulating spatial structure.

In string theory the importance of the horizon is recognized by the holographic principle, which states that the information of a volume of space is contained on the boundary (Susskind, 1994). This is particularly true for the information-saturated horizons of black holes. Taken together, the holographic principle and Landauer's principle mean that information accumulation of particles by the incessant standing-wave tick tock of the universe eventually uses up manifold energy (and macrodimensional volume), and turns those particles into black holes. As a consequence, the information-saturated black holes should be devoid of energy! This is congruent with recent investigation of entanglement near the horizon of black holes, which found that black holes are not more than their impenetrable horizons (Almheiri et al., 2015).

Interaction involves the formation of an **energy vacuum**. Energy vacuum is abhorred; therefore it is a temporary condition and the energy neutrality of the particle over space is restored by switching between standing waves in two, well-defined steps (Table I).

Step (1) Decreasing frequencies diverge and shorten to smooth the manifold, forming a **macrodimensional vacuum** (MaDV), which is a vacuum outside the **particle** (Figures 1.2.a). Like a spinning ice skater extending her arms to the side and bending her legs to reduce her length along the axis about which she is spinning, the shortened quantum wave cannot connect to the **spatial field** (SF). (The quantum energy, shaped as a wide and short diamond, is indicated at the center of Figure 1.2.a.) The MaDV therefore is transient and is instantaneously eliminated in step 2. Increasing frequencies converge as they expand and form a **microdimensional vacuum** (MiDV) inside the torus. Staying with the ice-skater analogy, she would pull in her arms and extend her body vertically in a straight line (**Figure 1.3**.a). (The narrow diamond-shaped quantum energy is indicated at the center of Figure 1.3.a.) The elongated quantum wave forms a strong connection with the SF, called down spin. Down spin expels MiDV, which travels on the SF as the **photon** wave packet and thus MiDV eliminated at an arbitrary spatial distance.

Step (2) The wave function collapses and reformulates on an altered frequency, and altered manifold energy. Increasing frequencies exchange

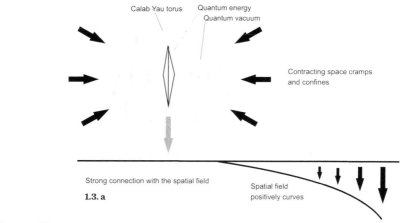

Calab-Yau torus · Quantum energy / Quantum vacuum

Contracting space cramps
and confines

Strong connection with the spatial field

1.3. a

Spatial field
positively curves

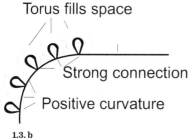

Torus fills space

Strong connection

Positive curvature

1.3. b

Figure 1.3. Structure of black holes
(a) Calabi–Yau torus at black holes. The arrows indicate the direction of energy inside and outside of the torus. The gray arrow shows the spin direction, the white arrows show the effect of the expanding manifold of the torus, and the black arrows designate the forces acting outside the torus.
(b) Structure of space at black holes. The weakened manifold of the Calabi–Yau tori stretches out to fill up space. The quantum wave forms strong connection with the SF, increasing its curvature. Their combined effects compact space. The lack of space turns quantum energy into shear stress, which is called the Weyl tensor.

Table I: Changes in energy of interacting fermions

Spin direction	Manifold	Quantum wave
Down Spin Field strength increases	- Information accumulates - The torus stretches out - Insulation is poor and discontinuous	- The frequencies increase - Convergence - The wave form elongates, interaction is encouraged - The torus volume increases - The degrees of freedom decreases
Up spin Field strength decreases	- Information is given up - Torus is well insulated and contracted - Space expands	- The frequencies decrease - Divergence - The wave rounds out, interaction is discouraged - The torus contacts - The degrees of freedom increases

macrodimensional space for horizon surface, while forming information memory and thus it is a dimension loss. Decreasing frequencies shrink the horizon surface, which contracts the torus, and forms macrodimensional volume (**Table I**). In this way interaction is a volume shift between the macro- and microdimensions.

As noticed, during interaction the volume of the micro- and macrodimensions changes inversely, which increases or decreases the dimensionality of space. As a consequence, the SF curves, just like inflating or deflating balloons attached to a surface would change the curvature of the surface (Figures 1.2.b and 1.3.b). This is analogous to the process in which two-dimensional silicon layers with varying tensions are attached (Xu et al., 2014). As the tension between the layers is released, the structure naturally acquires a three-dimensional shape. The curvature of the SF as a function of the information accumulation is shown in **Figure 1.4.** The dimensionality of space increases toward the left and decreases toward to right. The energy-information content of the horizon is indicated by the x axis. The two complex dimensions, space and time, formulate a four-dimensional universe. To remain congruent with the accepted directionality of time toward the black holes, the microdimension is designated as positive space, so a positive-curving SF forms gravity (Figures 1.1 and 1.4 galactic regions).

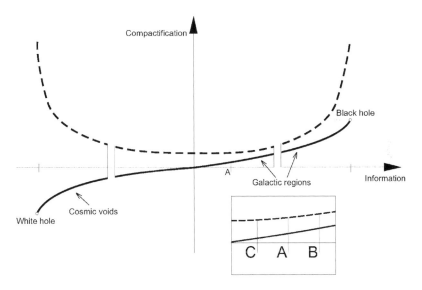

Figure 1.4. The four-dimensional universe The x axis is the information-energy ratio of the horizon that can be thought of as the age of the coordinate. The y axis represents the transformation of space from macro dimensions to microdimensions by compactification, thus it can only be theoretically traversed, by acceleration. In the following convention, microdimensions are designated as positive space. Therefore, white holes form at negative time and negative space (macrodimensions), whereas black holes form at positive time and positive space (microdimensions). Compactification changes the geometry of the horizon (the gravity-field curvature) according to the inverse sine function, as shown by the solid line. It is the limit of information transfer in the universe: the speed of light. The region above the curve is prohibited. Positive field curvature formulates the galactic regions, and negative field curvature is typical of cosmic voids. The curve's derivative (the Lorentz contraction equation) is the speed of the curvature change, represented by the dotted line. The dot (A) designates the totality of possible experiences of biological systems inhabiting the surface of a planet. The inset shows the total range of possible field curvature and temporal changes encountered due to minor changes of gravity. The dot (B) shows the effect of enhanced gravity. The compactification speeds up and slows the inner clock of material systems. Within smaller field strength, the speed of compactification is smaller, allowing clocks to run faster (C). However, the gravity field curvature (gravity) changes a lot faster than the speed of the clocks.

A fermion can be visualized as a single Calabi–Yau torus, but within the microdimensions the wave function is not restricted by space (or time). Our constrained macroworld view suggests a local character of space and a momentary experience of time. The „now and here" has the vertical, interconnected experience in the macroworld, whereas the particle wave function forms within the horizontal spatial infinitum of the microworld. In the microworld, the „here is there" and the „there is here," and standing waves are everywhere. It is pointless to ask where the particle is. It materializes where energetically convenient. In the oven the cake is heated wherever the cake is; in the same way, interaction occurs at the point where the wave function is disturbed. The point-like wave-function collapse (decoherence) binds the wave function to the here and now.

Euclidean spatial geometry is formed by equal concentrations of micro- and macrodimensions. The extent of spatial contraction as a function of information (time) is described by the inverse sine function (Figure 1.4). Intuitively, the curve forms the topological surface of space. Negative curvature forms cosmic voids, and positive curvature shapes galactic zones. The effect of negative curvature can be considered negative gravity (antigravity). The range of possible physical experience for inhabitants of a planet would be extremely narrow, as indicated by the dot (A). Although telescopes and particle accelerators can pry some physical characteristics of the first quadrant of the coordinate plane, the cold world of noninteracting **cosmic void**s would remain unapproachable. The derivative of the inverse sine function (Equation 1.1-1.3), which is the Lorentz transformation equation, expresses the fundamental, interdependent relationship between compactification and the information content of the horizon (age): space and time. Thus, the Lorentz equation can be considered as the speed of the change in the spatial angle of the topological surface of space between the poles of the universe. The relationship also predicts the degradation of spatial structure and the formation of singularities close to the poles (i.e., black and white holes):

$$\gamma = \frac{1}{\sqrt{1 - \frac{x^2}{c^2}}} \text{ (Equation 1.1)}$$

where c is the speed of light and x is the speed of the particle. Considering the speed of light to be unity simplifies the equation:

$$\gamma = \frac{1}{\sqrt{1 - x^2}} \text{ (Equation 1.2)}$$

The equation can be recognized as the derivative of the inverse sine function:

$$\frac{d}{dx}\sin^{-1}x = \frac{1}{\sqrt{1-x^2}} \text{ (Equation 1.3)}$$

Releasing information increases the manifold energy (up spin) and leads to negative time, whereas information increase corresponds to positive time. But the two conditions are connected. In the Feynman–Stueckelberg interpretation, particles are regular particles that are traveling backward in time (Stueckelberg, 1941). Therefore, the particle/antiparticle ratio is proportional to the rate at which the universe expands. In a contracting universe antiparticles would dominate. Age is thus the local information content of the manifold (e.g., the Calabi–Yau manifold), the holographic record (memory) of the spatial coordinate (Strominger et al., 1996). In accordance with Einstein's theory of relativity, the speed of clocks (not counting acceleration) is inversely proportional to gravity. Time begins in white holes; the high-energy manifold of the newborn Calabi–Yau torus lacks any information. Since the particle vibrations are low frequencies, the energy difference, and therefore the energy (information) exchange, in one **decoherence** is enormous. For this reason, clocks tick fastest there. In black holes the energy difference fades between the high-quantum-frequency overtones (the energy exchanged in one decoherence). Information saturation stops clocks and extends time to infinity (Figures 1.2.a and 1.3.a). Extreme positive spatial curvature compresses space into a point, but the information-saturated horizon expands. The positive spatial curvature forms pressure (Figures 1.3, **1.5**.e, and f). Due to the great density, force turns into shear stress (the **Weyl tensor**), which grinds up and agitates matter. In contrast, a new, energetic manifold is contracted, smooth, and nonreactant. However, the two conditions are symmetric and connected by the **Heisenberg uncertainty principle**. At white holes, time is newborn, but an expanding volume generates an uncertainty in position. At black holes, space contracts into a point, but time stretches out to infinity and becomes the source of uncertainty. Within Planck distances, gravity's constituents of MaDV and MiDV emerge as violent, untempered forces that equalize to form a stable force only on larger scales.

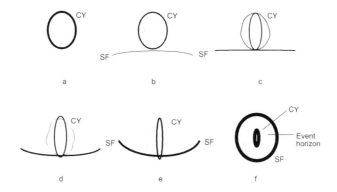

Figure 1.5. Connection possibilities of quantum energy with SF „CY" is the particle (a) Negative field curvature of a white hole. The quantum wave is nonexistent, so connection to the field is not possible. (b) Connection of quantum energy with negative-curving field. The quantum energy forms no connection with the field. (c) and (d) Quantum-energy connection with Euclidean field depends on both the local field and the spin direction of the particle, leading to spinor operation. Frequent decoherences turn the Calabi–Yau space into matter. (e) Greater frequencies elongate the quantum energy, which forms a strong connection with the positive-curving field. The manifold stretches out, whereas the positive-curving field squeezes, and space wanes. Quantum activity peaks but materializes as shear stress, called the Weyl tensor. (f) Positive field curvature of black hole surrounded by event horizon. The information-saturated manifold is wrinkled and stretched out. The quantum wave disconnects from the field and forms the event horizon. Notice that the quantum energy's connection to the field is proportional to the field's strength and that the angle formed by the quantum energy and the field changes between zero and 90 degrees.

Summary

The two-dimensional universe is formed by the orthogonal SF and TF. Since both SF and TF are complex dimensions, they form a four-dimensional universe. Time changes between energy and information, while spatial energy is formed by expansion and compactification. By changing in opposite ways during interaction, the SF and TF form energetic seesaws; an equal amounts of their constituents' energies forms the galactic, near-Euclidean environments. However, the polar ends of the universe are formed by only one component of either energy. Time begins in white holes, where MaDV is expanding space. In black holes, space is compactified into a point but time expands into infinity.

The Calabi–Yau torus forms standing waves, and their frequencies transform during interaction in a two-step process: (1) Contracting quantum waves form a microdimensional vacuum (MiDV), whereas relaxing quantum frequencies form a macrodimensional vacuum (MaDV). The MiDV travels on the field as a photon wave packet. (2) The universe abhors vacuum, so both the MaDV and MiDV are eliminated. The MaDV contracts the torus and forms space. The MiDV generates information, which weakens the manifold, decreasing the degrees of freedom. The information-laden torus expands. For outside observer the torus appears constant and non changing.

The Big Bang

The discovery of the cosmic microwave background radiation (CMBR) by Penzias & Wilson (1965) opened an important new phase of studies in Big Bang cosmology. The signal would permit the experiential study of the primordial universe.

Earlier, the innate relationship between time and the information content of the Calabi–Yau space was presented. Here, I postulate that this complex structure formed in the Big Bang and marks the beginning of time. Without Calabi–Yau space, there are no interactions. Time, gravity, temperature, and pressure depend on the existence of Calabi–Yau space, limiting the analysis and verification of primordial events. Nevertheless, the primordial energy of the universe is considered to be a vibrating string (**Figure 1.6**.a). Without mass (or gravity), neither pressure nor temperature could form, so the oscillations could occur with great abundance and fluidity. The initial conditions (the energy state of the string) would be unimportant, since the SF was transitory and structurally unstable. However, the amplitude of fluctuations would progressively diminish due to friction, as the string accumulated some form of spatial structure. Without the existence of time, the system is allowed to fine tune *itself* by evening out the amplitude of oscillations throughout the cosmos. The purely adiabatic initial conditions and nearly Gaussian distribution of the CMBR found in recent analysis supports this claim. The accumulation of kinetic energy by fluctuations of gradually decreasing amplitude would give rise to a constant energy potential of Calabi–Yau space, which separates space into a micro- as well as a macrodimensional domain, as shown by Figure 1.6.b (Candelas et al., 1985; Yau, 2009). Changes would come to require energy exchange between the different dimensions. The energy requirement of interaction would eliminate the oscillations, stabilizing the newly formed cosmos. *Therefore, the oscillatory energy of the universe accumulates as a potential and transforms into the self-energy of Calabi–Yau space.* The decreasing-amplitude oscillations would automatically smooth out the energy distribution over all space, independent of the path taken by the system. In preparation of dimensionality change, the system would slow down, which would allow fine tuning. The energy requiring simultaneous decoherence at the formation of Calabi–Yau space necessarily introduces small irregularities in the energy distribution. Like sand poured into an engine, the energy requirement of interaction

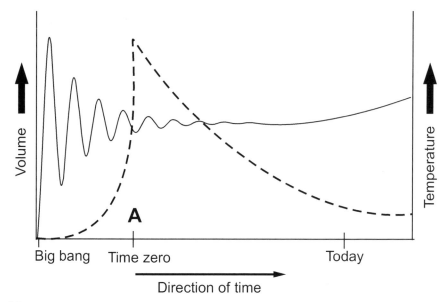

Big bang Time zero   Today

Direction of time

1.6. a

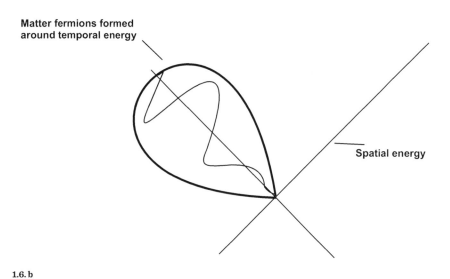

Matter fermions formed around temporal energy

Spatial energy

1.6. b

Figure 1.6. Cosmology (a) Volumetric and temperature changes of universe as a function of time. The volume of the universe is shown in solid line, whereas temperature is shown with a dotted line. Time did not exist immediately following the Big Bang; time keeping was launched with the formation of Calabi–Yau space. At the formation of Calabi–Yau space, time equals zero (A). (b) Formation of matter fermions limits interactions into quanta, which are the energy difference between standing-wave harmonics.

16

suddenly halted the primordial pulsations and „froze" the SF, stabilizing its almost infinitesimal gradation of energy. An *en masse* interaction at the formation of Calabi–Yau space could result in jamming, small irregularities in temperature as the universe heated up. The exhaustive generation of photons created a sharp spike in temperature and generated a uniform heat bath over all space, triggering recombination and nucleosynthesis. Scattering photons could generate small anisotropies and variations in temperature. The energy requirement of recombination and nucleosynthesis caused temperature to plummet (Challinor, 2012). The existence of the Calabi–Yau space introduced the energy requirement of interactions, which stabilized the SF. This simple mechanism fits well with many facets of the Planck data, such as the adiabatic, smooth, and scale-invariant fluctuations and the lack of gravitational waves (Ade et al., 2014, Ijjas et al., 2013, Ijjas et al., 2014). A recent three-dimensional map of the universe shows an unmistakable wavy, undulating distribution of matter (Carrick et al., 2015). Explaining the data without invoking primordial three dimensional waves, the result of string fluctuations is difficult.

Belinski, Khalatnikov, and Lifshitz (BKL) describe singularities in the cosmologic solution of the Einstein equations that have a complicated oscillatory character, causing time to be discontinuous. The proposed model also inevitably leads to polar singularities, which would direct the flow of time. The Big Bang is time zero for the universe. Time begins in the first discrete interactions between the micro- and macrodimensions. The smooth CMBR reveals the immensely equilibrated structure of space. Just as ice suddenly forms on the surface of an over-cooled pond, *en masse* decoherence would form Calabi–Yau space over all space, leading to the symmetric and smooth radiation density of the CMBR. Today, in contrast, photons are only freed from the surfaces of stars, which represent a mere fraction of the total energy density (Burgess, 2007).

The primordial fluctuations formed a loop, which could be the source of directionality in the universe, the basis of the half **spin** of **fermions**. Half spin is therefore a relic of the processes occurred prior the Big Bang. Since the TF, of the micro- and the macrodimensions of space are orthogonal, the interface of microdimensions and macrodimensions can flip (an 180^0 reversal of the directionality of the manifold) during interaction, a non-orientable connection. Thus, fermions form spinors.

SPATIAL EXPANSION IN THE AFTERMATH OF THE BIG BANG

The oscillations of the Big Bang formed considerable spatial volume, but the hot universe further expanded during post-Big-Bang cosmology. This spatial expansion can be envisioned as whipped cream that comes out of the tube as a thick fluid but expands and solidifies into foam. The formation of Calabi–Yau space (the beginning of time) turned a fluid, malleable energy field into a highly viscous, primordial space, but only the subsequent interactions formed the widespread network of gravitational fields, which solidified and expanded space into today's cellular, filamental structure. Thus gravity, temperature, and pressure are roughly proportional to age. The grainy structure (from which gravitational centers could form) of the CMBR already shows the origin of the universe's present complexity.

Summary

During the Big Bang the decreasing-amplitude oscillations of a vibrating string gave rise to the microdimensional Calabi–Yau space. The formation of Calabi–Yau space involved simultaneous decoherence over all space, which gave rise to low-energy irregularities and spiked temperature. The interaction of space and time came to require discrete pockets of energy. This energy requirement stabilized the structure of cosmos.

Connection to Theory of Relativity, Quantum Mechanics, and String Theory

Great scientists, such as Richard Feynman and Steve Wolfram, have suggested that the structure of the universe must be discrete (Wolfram, 2002). We have seen that wave-function collapse limits Calabi–Yau space energy levels into well-defined steps, and discrete magnetization, temperature, and gravitational structure is found even at the largest scales of the cosmos (Alfvén, 1981). The reconciliation of string theory, quantum mechanics, and general relativity is so prohibitive, as if they occupied different worlds. In a discrete universe this might actually be the case.

The pilot-wave approach to quantum theory, known as Bohm mechanics was pioneered by de Broglie (Bohm, 1952; Bell, 1987). The idea that a quantum wave, guided by a memory-retaining manifold, leads to quantum phenomena finds strong support from an unexpected source. Oil droplets guided by pilot waves in fluid experiments can recreate interference, tunneling, quantized energy levels, and other quantum phenomena (Brady & Anderson, 2014). Using weak measurements, trajectories of interfering photons can be reconstructed, which corresponds to results from pilot waves (Kocsis et al., 2011). The insulated microdimensions are energy-neutral; not affected by gravity, particle standing waves vibrate throughout the whole cosmos. The lack of spatial and temporal limitation permits **quantum entanglement** (Blaylock, 2010), the delayed-choice quantum eraser proposed by Scully and Druhl (1982), and all sorts of interference phenomena[2] (Megidish et al., 2013). Energy conservation demands that sister particles form one nonlocal energy unit with complementary wave functions, to be separated only by subsequent interaction. Thus, the drama of quantum mechanics and string theory takes place within the microdimensions of the TF, whereas gravity lives in the spatial macrodimensions. The three seemingly disparate theories (quantum mechanics, relativity, and string theory) converge exclusively through the singularity of the interaction between the torus and the field (the micro-and the macrodimensions) when the microdimensions materialize as a point-like scalar force (the Dirac delta function). Interaction is an orientation between contraction- and expansion, between micro- and macrodimensions, between lesser

---

[2] More detailed information on the background of several experimental findings regarding sister particles and entanglement can be found in the notes at the end of Chapter 1.

and greater degrees of freedom (Table 1). The rich gradation in torus volume due to expansion-compactification leads to the gravitational gradient. The probability equation of quantum mechanics originates in the Calabi–Yau structure. The energy potential and information content of the manifold are inversely proportional, so the manifold, quantum frequencies, and spatial curvature develop and change in concert by Lorentz transformations (**Figure 1.7**.d). The field influences the particle's energy level and in turn the particle changes the curvature of the field. Thus the energy level of elementary particles and the field curvature remain proportional in spite of constant changes and the elementary particle seeks out the field curvature that is congruent with its energy level. This was the prescient idea of Bayesian game theory (Harsanyi, 1968).

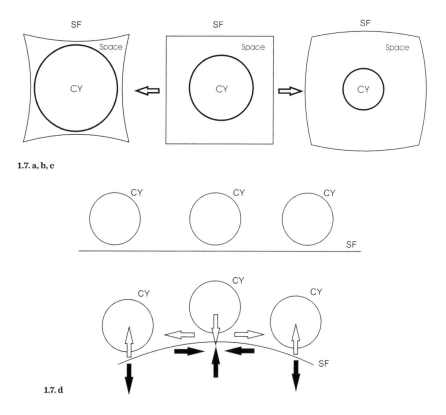

Figure 1.7. Possible spin states of Calabi–Yau torus (a)–(c) The Calabi–Yau torus is indicated by the central circles, and the surrounding squares represent the environment. (b) The central image represents the torus before interaction. (a) MaDV decreases in favor of MiDV in the down spin particle. The information increase expands the torus, which connects with the positive-curving SF. (c) In the up spin state the manifold contracts and expands space, initiating negative curvature. The shrinking torus cannot connect with the negative-curving SF. The actual volume changes are exaggerated for illustration. (d) A representation of the dynamic relationship between tori and the field curvature. The field curvature spreads within the environment and influences the energy balance of neighboring particles (white arrows). In turn, the neighboring particles influence the environment and the original particle (black arrows).

20

The structure of Calabi–Yau space makes quantum phenomena possible. Quantum mechanics and string theory are the consequence of microdimensions, whereas gravity emerges from the SF. The self-insulated, energy-neutral Calabi–Yau torus gives a discrete structure to space. Insulated from gravity, quantum energy enjoys complete spatial freedom. Interaction changes the standing-wave frequencies and also shapes the spatial curvature, a process that gives an undulating, curving structure to space. Entanglement is a common wave function of sister particles, which separate only upon subsequent interaction: up spin sacrifices quantum energy for space and, during down spin, the manifold records the information of the increasing quantum frequencies.

Spatial field

Moving between white holes and black holes, space is compactified into the microdimensions. Compactification, or positive time, parallels information accumulation and formulates elliptic geometry; see Figures 1.1 and 1.4 (galactic regions and black hole), whereas spatial expansion (negative time), leads to hyperbolic geometry; see Figures 1.1 (white hole) and 1.4 (white hole and cosmic voids).

Irregularities, which are the memories of past changes, elongate the increasing-frequency quantum wave, so it meets the field at an ever greater angle, forming more stable connections. The affinity or strength of such connections (Figures 1.5. a–e) reaches their limit at black holes, where the Weyl tensor fractures the quantum wave's connection, creating an event horizon (Figures 1.3 and 1.5.f). Although energy can turn into information (macrodimensions into microdimensions) and back, the basic structure of the universe is the same from black holes to white holes, irrespective of the distribution of mass! However, the universe's space-filling quality differs significantly, as expressed in the dimensions of energy ($J = Pa \cdot m^3$, where J is energy in Joules, Pa is pressure in Pascal and m is distance in meters). Near white holes, the limitlessly expanding space eliminates pressure, whereas in the environment of black holes, the space of the macrodimensions disappears. The microdimensions (positive space) create overwhelming pressure.

In moving between white holes and black holes, space is gradually transformed into microdimensional volume, while information accumulation weakens the manifold. Increasing quantum frequencies give rise to more elongated wave forms and increasingly stable connection with the stiff, curving SF. At black holes, the connection is severed by the Weyl tensor. Expansion in white holes is balanced by spatial contraction of black holes, yet an identical structure is retained over all space.

The anatomy of interaction

Fermions align themselves orthogonally to the SF (thus connected by Lorentz transformation), which makes interaction and energetic changes possible (Table I). The Pauli exclusion principle states that identical fermions cannot occupy the same quantum state simultaneously. In other words, within Euclidean fields, located in the middle of the curve (Figure 1.1, galactic region), only disjointed, contrasting changes can occur, which leads to the spinor operation of fermions and the Heisenberg uncertainty principle. The insulated Calabi-Yau torus determines discrete frequencies and discrete energy changes during interaction. The energy conservation dictates that interacting particles form complementary energy states. As we have seen earlier, particles form the spatial curvature of their coordinates, thus identical-spin neighbors would form abhorred energetic maxima (**Figure 1.8**.a). The stationary (minimal energy) formation constitutes the opposing-spin configuration as dictated by the Pauli exclusion principle (Figure 1.8.b). Interaction inversely changes the observer and the observed (up spin in one particle is necessarily balanced by the down spin of the partner particle), so energy and information changes are opposing and complementary. Because information accumulation (down spin) in one particle must be balanced by information release (up spin) from the other, even the smallest changes carry momentum. The contrary decoherences of spinor operation actually keep the field Euclidean! Within a curving field, temporal and spatial energies separate (Figure 1.1, and 1.4, black hole and white hole). Consequently, the particle is no longer a spinor, which obeys the Heisenberg uncertainty principle, but forms a deterministic behavior. The time-dilation equation clearly shows this fast-changing, curving region near the poles (Figure 1.4). Within gravitational regions, the direction of time is positive, which moves material systems closer to the high entropy of black holes. The positive spatial curvature brings

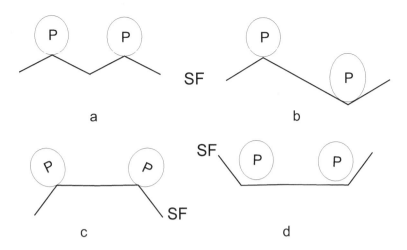

Figure 1.8. Geometric basis of Pauli exclusion principle „P" indicates a particle (fermion), „Field" indicates the curvature of the SF. (a) Two neighboring particles, both with down spin, form identical spatial curvatures, their connection form a local energy-maxima configuration. The same is true for neighboring up spin particles. (b) Two neighboring particles with opposite spins. The SF forms a minimal-energy configuration, as indicated by a straight line. (c) Two neighboring down spin particles within a positive-curving field represent the minimum-energy configuration. (d) Two neighboring up spin particles form minimum-energy configuration within a negative-curving field.

about a strong connection between Calabi–Yau tori and the SF. Such tori react to the slightest asymmetry of the field by increasing quantum frequencies and forming down spin decoherence (Figure 1.3), so that the manifold energy is eroded further still. Changes become self-perpetuating and irreversible. Within a positive-curvature field the minimum-energy configuration becomes the down spin (see Figure 1.8.c), and a 360° rotation restores the particle's wave function. At the other extreme, an enhanced manifold energy insulates the quantum wave from the SF, so the SF energy irregularities cannot reach the well-insulated quantum waves. As a consequence, only extreme quantum energy could produce down spin, but the negative curvature of the SF actually diverts and disperses quantum energy (Figure 1.2). Cosmic voids and white holes expand space and form negative time by erasing information. Up spin decoherence is encouraged, perpetuating manifold energy increase. Within the negative-curving field, energy minima are formed by the up spin configuration (Figure 1.8.d). Therefore, the down spin configuration is prohibited and a 360° rotation restores the fermion's wave function. An antiparticle's resistance to interaction reduces or completely eliminates the Pauli exclusion principle. Nonreactant **antimatter** falls apart, and the formation of baryonic antimatter is inhibited, as concluded by Canetti and colleagues (2012). At white holes, the temporal energy is zero and the spatial energy expands infinitely, whereas the positive-field curvature of black holes contracts space to a point.

The effect on interactions of changing field curvature can be envisioned as changing elevation affecting movement in physical space. Within a positive-curvature field, a ball moves down into a valley unidirectionally. However, by moving on the horizontal plane, the ball can easily change direction. Upon reaching the edge of the negative-curving field, the ball's movement will again unidirectionally follow the curvature downward.

The curving fields of the poles have a self-aggravating effect and limit the extent of the Euclidean regions between them.

The energetically opposite poles and their environments seem to be fundamental and necessary features of the universe. Interaction forms the field curvature, which translates into the pressure and temperature differences of the gravitational field. The aging of the universe must be balanced by information removal and expansion (a superfluous production of a virgin manifold), as shown by Moreva and colleagues (2013). Higher-manifold-energy areas with their negative spatial geometry form expanding vacuum bubbles, which repulse the path of everything, even light (a phenomenon opposite to that of gravitational lensing). The regions with higher manifold energy are energetically „younger" than the gravitational regions with their positive (contracting) spatial geometry. The rigid gravity areas retain a solid and stable large-scale structure, whereas smaller-curvature areas are highly flexible. Galaxies only form outside or around the regions of higher manifold energy, forming stiff structures of walls, filaments, or membranes. On the grandest scale, the cosmos is like a house of mirrors. Light converges due to the gravitational pull of large objects and diverges due to the negative-curving field. In regions of small field strength, interaction is inhibited. Quantum energy is deflected unchanged—a phenomenon known as conserved current. Surfaces surrounded by different entropic conditions react in an identical fashion, forming identical spin directions. For example, down spin surfaces (such as hot materials) lose energy due to their enhanced acceptance of decoherence.

Summary

Particles form the spatial curvature of their coordinates. Hence, the minimal-energy configuration exists only between opposite-spin particles, as stated in the Pauli exclusion principle. Only a 720° rotation restores the wave function of the spinor-vector particle. The positive-time evolution of gravitational environments is balanced by the up spin configuration of cosmic voids. Close to black holes the minimal-energy configuration becomes the positive field curvature, whereas negative field curvature is enforced around white holes. For this reason, close to the poles, a 360° rotation restores the particle wave function. The opposing energy changes within the positive- and negative-curvature regions form the energy neutrality of the universe. The regions rich in manifold energy are more flexible and younger than the positive-curvature gravitational areas, which form stiff galactic structures.

Mass

Einstein showed that mass can be turned into pure energy and that gravity has a geometric nature. Andrew Strominger and Edward Witten have surmised a connection between the torus and particle mass. The reduced Compton wavelength is the natural representation of mass on the quantum scale, which indicates the close relationship between mass and frequencies (particle quantum energy): $m = h/\lambda c$, where h is Planck's constant, λ is the wavelength of the photon, m is mass, and c is the speed of light in vacuum.

We have seen that the particle wave function and the field curvature are interdependent and that the eccentricity of the wave function is proportional to its connection with the field (Figures 1.5 and 1.8.a–c). The increasing information content means greater frequencies and increasingly deformed and elongated wave functions. Since the quantum energy meets the SF at an angle that is proportional to the information content (i.e., the age) of the particle, and black holes are deemed the heaviest objects, mass should be proportional to the strength and the angle of the connection between the quantum wave and the SF. For this reason, light speed can only be reached by a particle having no connection, therefore no mass. The connection of the particle and field is tightly regulated; the SF restores the field's original curvature against particles forming greater curvature. The vector force that updates the particle to the field corresponds to the particle's **weight**. Although the connection should be strongest near black holes, it gives way to the event horizon (Figures 1.3 and 1.5.f). At white holes, the energetic manifold is compact (the Calabi–Yau torus shrinks). The MaDV insulates the quantum wave, preventing any connection with the weakened SF. The grinding, agitating force of particle accelerators produces immense mass by degrading manifold energy.[3]

[3] The proposed breakdown mechanism of these unstable, high-mass particles is discussed later.

The possible differences in the organization of the Calabi–Yau torus would define the physical characteristics of fermions; namely, neutrinos, electrons, and quarks. The torus insulation focuses the intrinsic angular momentum of the particle toward a singularity, leading to its **charge**. The neutrinos lack electric charge, which could very well result from the total absence of the torus insulation. The small torus energy level of electrons would form the greatest electric charge: –1. The large Calabi–Yau torus energy could form a balanced wave function, which allows the fractional electric charge of quarks. We have seen that interaction inversely changes the manifold and quantum energy and the mass is proportional with the elongation of the wave function. As a consequence, increasing quantum energy should exponentially increase mass. Due to the shifting manifold quantum-energy ratio of the torus, the natural logarithm of particle masses of the three generations of fermions forms straight lines (**Figure 1.9**). The high-mass-forming spatial geometry of the weak nuclear force determines the range of standing-wave frequencies, which can form stable connection to the same field curvature. Each generation of particle family is thus defined by the range of standing frequencies. The vastly different mass members of the three particle families are stable only within the corresponding field strength of vastly divergent gravity environments. For example, the progressively increasing mass of muons and taus can be considered progressively older electrons. Within Earth's mild gravity, only the lowest mass of each particle type is stable. Neutrinos are known to change colors, and thus form exceptions.

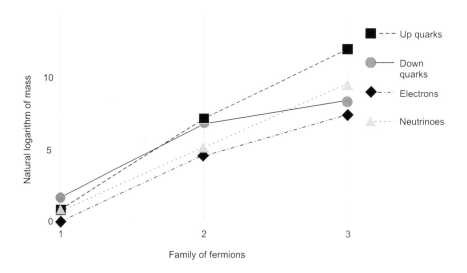

Figure 1.9. Natural logarithm of fermion masses The natural logarithm of fermion masses increases linearly within each particle type (quarks, electrons, and neutrinos). The only deviation is for down quarks.

Summary

The strength of the connection between quantum waves and the SF defines mass. The SF maintains its curvature against the local particle by a force that corresponds to the particle weight. The three existing particle groups have vastly different torus structures. Neutrinos, electrons, and quarks display progressively increasing torus energy levels. Within each particle group exists three particle families with increasing masses (decreasing manifold and quantum-wave ratio). Because the manifold and quantum energies change inversely to each other, the more elongated wave function of particles with greater quantum energy forms exponentially greater mass. Indeed, plotting the natural logarithm of particle mass produces a straight line. Only the particles with the smallest mass (greatest manifold ratio) from each particle family are stable on Earth.

Entropy within the universe

"Entropy measures lack of information; it also measures information. These two conceptions are complementary"

– Brissaud 2005

The concept of entropy originated in the nineteenth century to distinguish the thermodynamically unusable energy of the system. It is a law rooted in experience and experimental data, but it has remained a difficult, confusing, and conflicting definition, making entropy the most attacked and defended concept of physics. Many fields (such as information theory or quantum mechanics) have their own, although related, definitions (Duncan & Semura, 2004; Martin et al., 2013). Beginning in the 1980s, the second law of thermodynamics has been increasingly challenged (Capek & Sheehan, 2005). Even great scientists such as Stephen Hawking (1988) emphasized its limited, statistical power. An early contradiction of the second law was Poincare's recurrence theorem, but more recent experiments challenge the law's very foundation by questioning its validity within small time scales and small distances (Wang et al., 2002; Gieseler et al., 2014; Murch, 2015).

Intuitively, the above arguments have shown that entropy can be viewed as the extent, or strength, of the connection between the quantum wave and the manifold energy. A strong connection impedes interaction and corresponds to the stability of high entropy.

In low-entropy conditions, the manifold and the quantum wave easily separate to form MiDV (energy available for work)—the boson. Entropy is conserved and changes inversely in interacting systems. This idea is a direct challenge of the second law of thermodynamics. The high-entropy regions of the universe are the poles (**Figure 1.10**). The vicinity of the expanding white holes lacks quantum energy (Figure 1.2.a) and the weakened manifold of the contracted black hole regions cannot separate from the overactive (energetic) quantum wave (Figure 1.3.a). This is in line with Brissaud's idea. If we look at entropy in terms of information (a very traditional approach), then white holes clearly have high entropy because they lack information, whereas black holes are information saturated. Half way between the poles, the midpoint in the information level generates the greatest complexity near Euclidean spatial field. The black holes are quantum, whereas the white holes are manifold energy strongholds. Their opposing effects maintain the low entropy of the galactic regions, causing interaction—the dynamic, endless, and reversible transformation between the quantum and manifold energies. Inadvertently, interaction forms gravity, which binds these regions together. At the galactic regions, entropy is a relative state. High entropy occurs if the quantum energy is smaller than the local entropic balance, or if the manifold energy is greater than the local entropic balance. In low-entropy situations, either the quantum energy is greater than the local entropic balance, or the manifold energy is weaker than the local energy balance. Manifold energy (and space) is the energy reserve of the universe, but only through conversion into quantum energy can it turn into work.

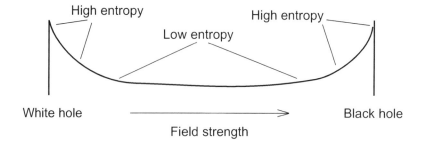

Figure 1.10. Entropic changes within the cosmos Poles (both white holes and black holes) display high entropy, in spite of their contrary energetic structures. Low entropy occurs at the near-Euclidean fields of the galactic regions. The SF strength increases between the poles, the small field strength near the white holes reaches its maximum at the black holes. White holes represent the high entropy of the manifold energy that lacks information. Near black holes, high entropy is due to the information-saturated manifold. The tension between the poles instigates constant decoherence within the low-entropy Euclidean environments, where fermions undergo spinor operation. Unceasing interaction is the source of sensory experience (i. e., matter), which is typical of the low-entropy galactic regions.

Experience and **matter** depend on decoherence. Constant interaction turns a spot in the sky into the yellow of the moon and turns a caress into a feeling of softness. Without interaction, the galactic region would just be a sea of energy. Decoherence is asymptotically inhibited at black holes (although their surroundings display the greatest activity in the universe) and, near white holes, no quantum activity exists that would allow interaction. The poles are not open for experience; their existence can only be inferred. In our expanding universe, the energy of expansion is greater than that of contraction, so the balance constantly tilts toward expansion, which is a tendency for up spin decoherence. On Earth (as on every gravitational body), the effect of gravity reverses this tendency and leads to the slow, irreversible breakdown of material systems. The tendency for increases in quantum energy within regions of gravity is formulated by the second law of thermodynamics. In low-gravity space, the propensity for entropy decrease can lead to greater-than-expected acceleration (Figure 1.4). Inhibition of decoherence (and matter) gives rise to cold cosmic voids near the negatively curving field of white holes. Expansion decreases the entropy of the universe's constituents, ensuring the renewal of material systems.

Summary

Entropy is the extent of the connection between the quantum wave and the manifold energy. A strong connection corresponds to high entropy, whereas a weak connection means small entropy. In low-entropy conditions, the quantum wave easily separates from the manifold and forms a MiDV—the boson. Interacting systems conserve entropy by undergoing opposite entropic changes. High entropy occurs near the poles of the universe. Near black holes, manifold energy is too weak to separate from quantum energy. Near white holes, the lack of quantum energy ensures high entropy. Low entropy is limited to the near Euclidean galactic regions. The tendency for down spin decoherence in gravitational regions leads to the second law of thermodynamics.

Elementary Forces

Gravity

Newton's universal law of gravitation and the Wheeler-DeWitt equation lack time dependence. Gravity, an overflow of the contraction force, does not depend on time. Instead, constant decoherence gives birth to time and spatial complexity by reallocating macro- and microdimensional space. Gravity is the resulting positive spatial curvature. When examined localized, decoherence appears to be an arbitrary and chaotic process, in which interaction drives systems toward equilibrium (Linden et al., 2014). Einstein's equation describing free-falling particles shows that pressure causes gravitational attraction. Down spin decoherence is more likely within a field with slightly enhanced pressure. In the up spin entangled pair, MaDV expands space and decreases spatial curvature and pressure (decreasing gravitational attraction). Within the most frequent decoherence areas, pressure and MiDV accumulates, which engenders down spin and enhances positive spatial curvature. An orderly chain of events gradually increases spatial curvature in the geometric center and channels space outward, decreasing spatial curvature (**Figure 1.11**). The contrary evolution of microdimensions (torus volume) and macrodimensions creates a tension, which bends and curves space into the fourth dimension, as discussed in the section „The Heuristic foundation of hypothesis" (Figure 1.7.a,b,c). Within regions of greater curvature, the particle horizon stretches out (rigid, information-saturated manifold), but the macrodimensions contract (Figure 1.3). In the center of large bodies where curvature is greatest (a planet, or star for example), pressure accumulates, leading to rigidity and higher temperature (Figures 1.4 and 1.11). Gravitational contraction forms the tight and stiff presence of material aggregates, such as galactic objects. Based on the difference between the speed of clocks and the value of the gravitational acceleration (the Lorentz equation and inverse sine function), the field curvature can be calculated.

The positive spatial curvature is not limited to areas of decoherence; it also shapes interplanetary, or intergalactic space. Just like an emperor with his courtiers, planets and galaxies move with their enormous positive-field-curvature envelopes, which are invisible and resist measurement. The congruent evolution of the particle and the field curvature would permit an unexpected access to information, as recognized by Harsanyi

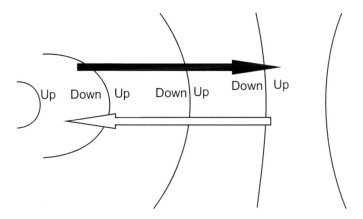

Figure 1.11. Gravity Curved lines represent different spatial curvatures in large objects. The Pauli exclusion principle forces interaction between neighboring particles. The greater spatial curvature forces down spin on the particle sitting within its perimeter; whereas up spin decoherence is encouraged in particles within less spatial curvature. The word pair „up" and „down" indicate the *probable* spin directions of the neighboring particles. The black arrow shows the direction of a MaDV drifting toward less spatial curvature. The white arrow shows the flow of a MiDV (torus volume—information accumulation) toward the central, increasing-spatial-curvature region of the gravitational body.

(1968). Based on Bayesian game theory, he claimed that nature (the underlying field) is also a participant in the game (interaction). Thus, gravity is mediated by the pervasive presence of the SF curvature. Like court events arranged ahead by servants, the positive curvature of the SF traverses space to prepare stellar meetings in advance. As two or more field curvatures superimpose, neutral gravity gradients can form Lagrangian points as the common field moves and rearranges matter up to the largest scales. Verification of this proposition would mean that the Pauli exclusion principle and the resulting entanglement fully describe gravity—the need for a special particle (graviton) becomes redundant.

The current understanding of gravity is challenged by, among other things, the flyby anomalies (gravity-assist maneuvers) and the formation and behavior of galaxies, such as the faster-than-expected speeds of stars at the outskirts of galaxies. The baryonic matter of galaxies seems to be embedded within halos (**dark matter**) of gravitationally bound structure (Frenk & White 2012). The close relationship between the galaxy mass (which is largely dark matter and so has no luminosity) and the intrinsic luminosity of a galaxy, called the Tully–Fisher relation, is difficult to interpret without considering the effects of spatial curvature. The white holes within antimatter nurseries of cosmic voids form expansionary bubbles (Figures 1.4 and 1.7). The expanding space pulling on the rest of the cosmos can produce the appearance of **dark energy** (Iacovlenco, 2012).

Gravity appears to be an attractive force, but it is a consequence of pressure coming from the expansion in cosmic voids. As an opening pivot, spatial expansion enhances pressure (*appearance* of mass) within gravitational regions, all the way to the center of galaxies (Figures 1.4 and 1.7). Consequently, the accumulation of mass is found to follow a power-law distribution, in accordance with the curving SF (Figure 1.1). Similarly, the enhanced pressure of SF curvature can stimulate greater activity, which increases the temperature and luminosity of galaxies (in agreement with the Tully–Fisher relation). The minute deviation in the flyby speed of satellites could also result from spatial curvature. As the satellite accelerates (or decelerates) between different curvature regions, the frequency of decoherence is modulated to produce deviations from expected values. On the largest scales, the behavior of the universe seems to be dictated by spatial curvature.

Summary

On Earth, as on every gravitational body, the gravity-field curvature creates the experience of gravity. Neighboring particles form opposite spin configuration, as dictated by the Pauli exclusion principle. Down spin states create the greater spatial confinement of positive spatial curvature, which increases the likelihood of down spin. Within regions of greater curvature, interaction occurs with greater frequency, exaggerating the differences in field curvature between the center, which has the greatest curvature, and outlying regions with less frequent interaction. The outcome of interaction becomes dependent on its location within the onion-like energy structure of the field, which directs material movement even on the grandest scale. Lagrangian points with neutral gravity gradients can form at the intersection of field curvatures. Bayesian game theory recognizes the importance of the role of the field in directing interactions. This hypothesis explains gravity without requiring a special particle.

The expansion caused by the white holes can manifest as dark energy. The proportional relationship between the luminosity and rotational velocity (the function of mass) of galaxies suggests the existence of extra matter, challenging our understanding of gravity. However, expansion can create pressure in symmetric, positive-curving gravitational regions, causing the appearance of extra mass, which is dubbed „dark matter." The movement and behavior of galaxies, as well as the flyby speed of satellites, could be directed by the spatial curvature, causing a deviation from calculations.

RELATIONSHIP BETWEEN SPRING MOTION AND GRAVITY

Interaction redistributes space between the micro- and macrodimensions, which curves space into the fourth dimension and give rise to the cellular structure of cosmos. Quantum energy (MiDV) causes constant activity (such as heat and pressure) in the center of large bodies, whereas expansion forms cold, cosmic voids. The equal balance of quantum and manifold energies within Euclidean areas breeds flexibility and fast-paced change. The interaction of the particles formulates the field, and the field determines the behavior of the elements from which it is made. Spring motion seems rather different. Springs oscillate in simple harmonic motion, propelled by a restoring force that is proportional to the displacement. The spring movement gradually slows and comes to a complete stop at the equilibrium point, as described by Hooke's law (**Figure 1.12**.a)

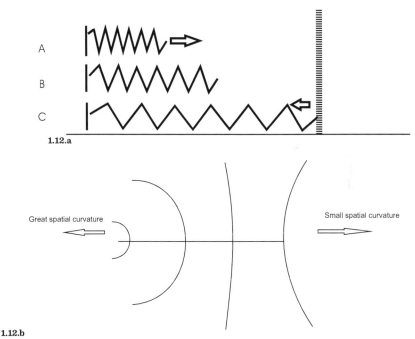

1.12.a

1.12.b

Figure 1.12. Analogous energy structure of simple harmonic oscillation and gravity (a) Both the contracted (A) and stretched (C) spring create tension, and the spring moves toward the equilibrium position in both cases. The arrows indicate the direction of the restoring force. The tension is maximal when the speed is minimal (zero). At the equilibrium position the speed is maximal, but the restoring force is zero. The position and speed are orthogonal. (b) A schematic representation of the changing spatial curvature of a gravitational body. At the far left, quantum energy is maximal and the field curvature is greatest, which squeezes space. At the far right, quantum energy is minimal (zero), and space is generous. The event frequency (decoherence frequency) and the transformed energy are inversely related. The energy transfer is greatest when the quantum and manifold energies (MaDV and MiDV) are in equilibrium (at the Euclidean curvature field, which is the geometric center of the figure). MiDV is forced toward positive spatial curvature (left-pointing arrow), whereas MaDV gravitates toward negative spatial curvature (right-pointing arrow). At both extremes, the energy transfer is reduced to zero. The SF curvature is analogous to the tension of the spring, and the event (i.e., interaction frequency) and spring frequencies are analogous.

F = $-kx$ (Equation 1.4),

where F is the restoring force, k is the spring constant, and x is the displacement from the center. Because quantum and manifold energies obey the Pythagorean identity, the derivation is satisfied by the trigonometric functions: $-\sin x$ (quantum frequencies) $\rightarrow -\cos x$ (manifold energy) $\rightarrow \sin x$, where „$\sin x$" is the field strength (the derivation of the **gravity gradient tensor** from the gravitational potential). This way, MiDV turns into the horizon surface and finally into MaDV. Thus, the micro- and macrodimensional spaces are the inverses of each other! This relationship illustrates the inherent connection already discussed between quantum frequencies and field strength (Bayesian game, Harsanyi, 1967 and 1968).

If the event frequency in gravitational relationships is considered analogous to the frequency of harmonic motion, and the tension of the spring is viewed as the tension of SF curvature, then the similarities between the two phenomena become clear (Figure 1.12). At the greatest deviation, the spring movement comes to a stop and interaction ceases. At white holes the energy transfer of one decoherence is immense, but the frequency of interaction is zero (**Figure 1.13**). At black holes the frequency of decoherence is maximal, but the energy transfer decreases to zero. The speed of the spring is greatest at the equilibrium position and the energy transfer is greatest at Euclidean regions. However, in contrast to the decreasing amplitudes of spring motion, interaction increases the

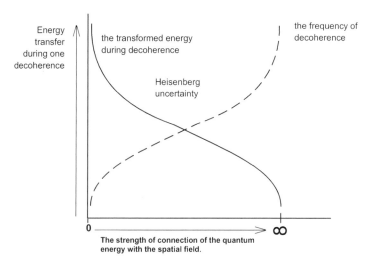

Figure 1.13. Energy transformation of decoherence between white holes and black holes The white hole state is 0, the black hole state is ∞. The field strength and its connection to the quantum wave increases from zero (0) to infinity (∞). The amount of energy transformation in one decoherence and the frequency of decoherence are indicated by a continuous line and a dashed line, respectively.

differences in field curvature and moves the system toward its poles! By representing the energy transfer against the SF curvature it would form a bell-shaped distribution. If the amplitudes of harmonic motion were registered over space, the decreasing amplitude of oscillations would also form a bell curve, unveiling the analogous energy structure between harmonic oscillations and gravitational field. The gravitational gradient forms the global shape of space according to the equation of harmonic motion (Equation 1.4). Thus, the constant k determines the shape of the curve and the behavior of the system.

Thus, the global shape of the cosmos can be viewed as a spring that is continuously being expanded by the Pauli exclusion principle (the principle of stationary action), and the negative field potential of expansion and the contracting force of the gravitational regions should be in equilibrium, as shown earlier (Moreva et al., 2013). The violations found of the second law of thermodynamics (Wang et al., 2002; Gieseler et al., 2014) support this notion.

Summary

Spring motion is described by Hooke's law. Gravity forms increasing differences in field curvature, where the quantum energy (torus volume) accumulates within the center, and manifold energy (space) accumulates within the layers with lesser curvature. Within the layers with Euclidean curvature, the quantum and manifold energies are in equal balance. If the SF curvature were represented over their distance from the equilibrium, a bell-shaped distribution would indicate an analogue energy structure between harmonic oscillations and the gravitational field. Harmonic motion forms a bell curve of decreasing amplitudes, whereas interaction increases the spatial-curvature differences. The frequency of harmonic motion equals the event frequency of gravitational relationships, and the increase in the SF curvature corresponds to the tension of the spring.

ACCELERATION Particles form the most stable connections with a slightly smaller SF curvature, which causes simple, garden-variety gravitational aging due to the field's constant attempt to restore its curvature (by forming the particle weight). However, far greater spatial curvature pushes particles toward smaller field strength in a process called **acceleration**. Although both gravity and acceleration are energetic moves toward higher entropy and both slow the inner clocks of material systems, there are opposite reasons behind them.

Positive SF curvature is generally called gravity. The concerted aging of gravitational regions is enshrined in the second law of thermodynamics. The continuous erosion of manifold energy slowly increases the SF curvature, so the flow of time is positive. Down spin decoherence increases the information content of the Calabi–Yau manifold and decreases differences between quantum-frequency overtones. The decreasing energy transformation during one decoherence slows clocks, in agreement with general relativity. Thus, the speed of the inner clock is inversely proportional to the field strength (**Figure 1.14**.a and 1.4, inset).

Acceleration reverses the gravitational process. When a particle is subjected to excessive gravity, then the particle disconnects from the field and the SF propels it toward lesser curvature (Figure 1.14.b). This so-called acceleration corresponds to up spin decoherence: the particle's energetic move toward a region with less spatial curvature. The Calabi–Yau manifold energy increases, flowing time backward and decreasing the frequency of decoherence. In everyday practice, the greater field curvature is imitated by compression. Acceleration is limited to near-Euclidean environments, because the field with negative spatial curvature is weak and extinguishes the possibility of interaction. In contrast, the pressure of immense field curvature does not have sufficient volumetric flexibility for acceleration. Acceleration shrinks the size of particles but expands the space between them. As a result, accelerating particles display great spatial flexibility and tolerate compression quite well. The decrease of the field curvature corresponds to information release (in agreement with Landauer's principle), leading to the Unruh temperature (Unruh, 1992).

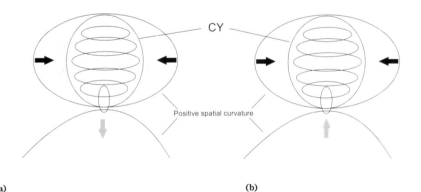

(a)   (b)

Figure 1.14. Gravity and acceleration of particles CY indicates the Calabi–Yau torus. Arrows indicate energy directions. Spin direction is shown by gray arrows. Black arrows indicate the pressure of positive spatial curvature. White arrows show the change in particle volume. (a) The manifold of the down spin particle expands, forming down spin decoherence (see white arrows). (b) During acceleration, the manifold contracts (shown by white arrows) and forms up spin decoherence (gray arrow).

Summary

When the strength of the SF and a particle's own curvature are congruent, they form a stable gravitational connection. As the energy difference between the particle's harmonic frequencies decreases, their inner clocks slow. The opposite process gives rise to acceleration. Within excess gravitational confinement, the particle disconnects from the field; the field's own strength accelerates the particle toward a region of lower SF curvature. In everyday practice, excessive gravity is imitated by compression. Acceleration corresponds to negative time and less frequent interactions, slowing the inner clocks of particles.

Electromagnetic force

The fundamental electromagnetic force is encountered frequently in material interactions. It causes oppositely charged particles to attract and identically charged particles to repel. Like gravity, the electromagnetic field is inversely proportional to the square of the distance between objects and spreads without limit, indicating its connection to the SF. Fermions are tiny bar magnets, which orient along compactification and expansion potential of the field (**Figure 1.15**). Down spin particles lose dimensionality, which exposes their wave function (i.e., charge) toward the field. In contrast, up spin particles gain dimensionality in the direction of the field, which insulates their wave function. Thus an electron is oriented between the differences in dimensions (spatial expansion-contraction potential) of the spatial field. For example, a down spin electron loses dimensionality by compactification, but an up spin electron expands and gains dimensionality as it turns toward the field. However, we never notice this orientation, because the far-stronger electromagnetic force overpowers it and determines the electron's orientation. The Pauli exclusion principle permits spatial proximity only for opposite-spin particles. The exposed quantum energy of down spin particles enhances the local field strength (Figures 1.5.d,e, and 1.15.a), whereas the well-insulated up spin neighbors weaken it (Figures 1.5.a, b, and 1.15.b). The difference in field strength orients the two particles. The tension between down spin and up spin states, or between regions of positive and negative spatial curvature, forms an orthogonal energy flow. In these energy pulsations, regions with down spin conditions become energy starved whereas regions with up spin conditions become energy rich, which is the source of periodic harmonic motion. The energy flow follows the angular momentum of the originating down spin particle.

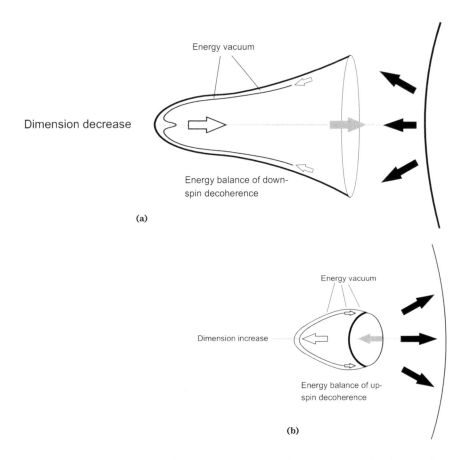

Figure 1.15. **Energy structure of fermions** The arrows indicate the direction of energy within the torus. White arrows show the quantum energy flow within the torus. Gray arrow shows the spin direction and black arrows show the particle's influence on the surrounding field curvature. (a) In down spin particles the converging quantum waves elongate, forming strong connection with the SF, which enhances field strength and the tendency for interaction. (b) For up spin particles the energy flow is divergent (traces the manifold), forming weak connection with the SF. The MaDV insulates the particle from the SF and discourages interaction.

Opposite-spin particles attract and identical-spin particles repel. The energy is constant during movement, although its direction constantly changes, as directed by the curving field (**Figure 1.16**). Both electromagnetism and gravity rely on the particle spin state and the field curvature, leading to similarities in their effects.

Magnetic fields appear to be part of the galactic structure and their presence on many scales of the universe challenges assumptions (Widrow, 2002). However, the three-dimensional map of the universe shows that distortions of the spatial structure by primordial fluctuations have survived the Big Bang and influenced the distribution of galactic matter (Carrick et al., 2015) and extragalactic magnetic fields. Because electromagnetism, like gravity, originates from uninsulated quantum

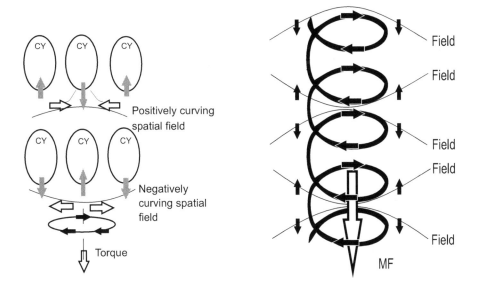

Figure 1.16. **Electromagnetic force** „CY" indicates particles, „Field" is the SF, „MF" is the magnetic field. (a) Gray arrows indicate the spin direction of particles. Upward arrows indicate up spin, downward arrows indicate down spin. For the down spin direction, the intrinsic angular momentum is exposed and becomes the basis of the electric field. The particles in the surrounding up spin manifold act as insulation and form magnetic flow. The exact details of the mechanism are as follows: the particles enhance the field curvature of the local field according to their spin direction. The restoring force of the field acting against the particles generates a tension that leads to the flow of electric charge. (b) Formation of electromagnetic field. The white arrow shows the direction in which charge flows, which follows the spin direction of the original down spin particle. The black arrows show the field restoring force. The two combined effects generate the magnetic field (black curving arrows).

energy, it is proportional to field strength. As a consequence, within areas of greater field strength, gravity becomes a greater force relative to electromagnetism. In a celestial body the differences in gravitational field curvature would necessarily generate an outward electric flow (**Figure 1.17**). This minute flow would remain constant over the surface, making it impossible to detect. A noninteracting torus is not compatible with an electromagnetic field, so magnetic field lines form at the boundary. For a large, spinning gravitational body, the deviation in the speed of rotation between different layers could generate a dynamo effect that is similar, for example, to the observed magnetic field of planets or stars. The same mechanism could help shape the large-scale magnetic fields of galaxies.

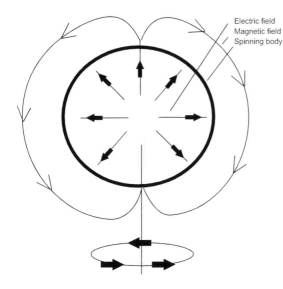

Electric field
Magnetic field
Spinning body

Figure 1.17. Formation of electromagnetic field in spinning gravitational body
Arrows indicate the direction of the flow of charge. The electric potential of a particle is greatest in the gravitational center of the body and generates an outward flow of electric charge (as indicated by black arrows within the spinning body). The axis of the spinning body becomes the exit and entry positions for the magnetic field, which surrounds the object.

Summary

Fermions orient themselves against the curving SF. In the down spin state the exposed quantum energy can connect with the SF, increasing its curvature. In contrast, the up spin state enhances the torus insulation and weakens the local field strength. As a result, opposite-spin particles form an asymmetric field curvature between them, creating a tension between push and pull. The asymmetry of the field over large scale gives rise to the electric and magnetic fields of galactic objects.

The weak nuclear force

The weak nuclear force is a gravity-adjusting force that assists in breaking down fermions and changing their charge to maintain charge neutrality throughout. The weak nuclear force saves fermions from a black hole fate and gives a second chance to stellar matter, which inadvertently conserves the quantity of matter in the universe and maintains the smooth topology of SF. In contrast to the long and even unlimited range of gravity and electromagnetism, the weak nuclear force has a notably short range, indicating its connection to the torus. As indicated earlier, particle mass has a geometric nature. Down spin elongates the quantum wave, and increases the information content of the manifold. Thus, particle accelerators „age" particles by expanding energy (Figure 1.9). Since the elongated wave functions of high-mass particles cannot integrate into the

SF of Earth's slight gravity, they have a short lifetime. The weak nuclear force inverts the curvature of the particles' field. The change of the field curvature also affects the torus by shifting the standing-wave frequencies (the manifold-quantum energy ratio). The breakdown mechanism by the weak nuclear force consists of two, well-separable steps: In the first step, the weak nuclear force inverts the field curvature, forming a negative field and giving rise to an unstable semi-white hole torus (Figures 1.5.a and 1.15.b). In the second step, the Euclidean field ages the white hole Calabi–Yau torus by moving it forward very fast in time and space (beta radiation). Frequent interactions rapidly age the particle, until its field curvature is equilibrated with that of the environment. In quantum field theory the vacuum is purported to fluctuate with activity, bringing virtual particles in and out of existence. However, under circumstances involving great energy, the aging of particles and corresponding curving field triggers the weak nuclear force, which would restore the SF curvature. The sudden changes in field curvature involving energetically highly turbulent dimensionality modifications would be experienced as quantum fluctuations. On Earth, only the lowest-mass particle from each particle family is stable but, in all likelihood, high-mass particles are abundant within regions of immense gravity, such as in the dense interior of neutron stars. The weak nuclear force only affects high-gravity (contraction energy) particles and only expansion-energy antiparticles; a phenomenon called parity violation in physics.

Summary

Only particles of compatible mass can connect to the SF. Particle accelerators increase particle mass by degrading manifold energy. These particles with enhanced quantum energy cannot connect to the local SF, leaving their immense field curvature unstable. The weak nuclear force inverts the particle's field curvature, which decreases the manifold-information content, creating an unstable semi-white hole state. Local environmental effects age the particle quickly, causing it to adopt the spatial curvature of its environment: an appropriate rearrangement of the manifold quantum energy ratio.

The strong nuclear force The strong nuclear force is the strongest elementary force and confines quarks within hadrons.

Conclusions

Space, which is energy and time (the latter of which is information) can transform into each other through interaction.

This chapter proposes a hypothesis with a new vision for the Big Bang and an operational principle of elementary-particle behavior. The main features of this hypothesis can be summarized as follows:

1. The fluctuations of the Big Bang formed microdimensions of the TF called the Calabi–Yau space, limiting oscillations into standing waves. The TF and the macrodimensional SF are orthogonal and form a highly symmetric universe.

2. Time, pressure, and temperature each rely on Calabi–Yau space; thus, each began in the Big Bang. The CMBR is the glow of the first decoherence, which was *en masse* and which led to a sharp spike in temperature.

3. The horizon of Calabi–Yau space separates two parallel worlds with sharply contrasting features. Inside the torus, microdimensional quantum waves form a universal expanse, but the macrodimensional SF exhibits local features. Interaction is the collapse of a wave function, forming discrete, point-like energy states. The informational record of decoherence is time. We see time as the cause of events, but in fact decoherence generates time, which is a record of the transformation between volume (negative space that reduces pressure) and information (positive space that forms pressure).

4. To understand our world, we rely on sensory experience generated by the decoherences of the near-Euclidean and positive-curvature regions of material systems. The principle of stationary action (the Pauli exclusion principle) forms gravity by increasing differences in the curvature of the field, leading to the non-Euclidean boundaries (poles) of the universe, which are hidden from experience. The spatial curvature attracts particles with greater spatial curvature, but repels particles with smaller curvature (forming acceleration).

5. The immense energy of particle accelerators erodes manifold energy

and curves space, forming particles with immense mass. However, the weak nuclear force inverts the SF curvature and turns these particles into expansion energy (white hole states). The white hole particles expand the cosmos by producing new Calabi–Yau tori (beta radiation). Thus, accelerator experiments verify our understanding of the Big Bang, the structure of black holes and white holes, and the operational mechanism of the weak nuclear force.

6. This hypothesis turns quantum mechanics into a comprehensible science. Quantum mechanics and string theory emerge from the microdimensions, whereas gravity is born from the SF. Within gravitational environments, the tendency for down spin decoherence leads to the second law of thermodynamics (the aging of matter). The rate of the universe's expansion determines the matter-to-antimatter ratio. The lowest-energy configuration of neighboring particles is for them to have opposite spin, forming entanglement. Thus, the Pauli exclusion principle, a time-tested principle of physics, naturally emerges from the principle of least action. Eventually, interaction seals the particles' polarized wave function and engenders gravity.

This hypothesis is consistent with existing observations and knowledge and predicts future observations or experiments that can support it or disprove it. The difference in the gravitational field strength should lead to a small but discernible difference in the second law of thermodynamics in space and on Earth and taking a longer time to reach an equilibrium. Therefore, entropy is expected to increase less in space than on Earth. When subjected to immense gravity, particle orientation against the SF (dimension change) could produce a measurable, albeit tiny, electric field. Finally, gravity and acceleration (down- and up spin decoherence, respectively) would change the volume of objects in opposite ways—a difficult measurement with current technology. Alternatively, the relationship between pressure and the speed of inner clocks would indicate gravitational effects.

Finally, this hypothesis defines time as the information content of the manifold and defines entropy as the strength of the connection between the quantum and manifold energies. The increasing entropy of gravitational environments is balanced by the decreasing entropy outside of these environments. It recognizes the geometric nature of mass and the source of weight. It also offers a deeper understanding of the fundamental interactions and gives a reason for parity violation of weak interactions. The decreasing entropy within the universe is an inexhaustible source of energy for the formation of new worlds and galaxies.

The hypothesis forms a coherent physical world view based on the tested principles and laws of physics without requiring arbitrary, adjustable elements or introducing exotic particles. I would welcome suggestions, criticisms, corrections, and collaboration from interested physicists, mathematicians, or other scientists for further development and testing of these ideas.

Notes

Details of quantum mechanical phenomena and experiments are summerized below.

SCHRÖDINGER'S CAT Schrödinger's famous thought experiment takes a cat as a stand-in for a quantum particle. Schrödinger's cat, just like a particle, seems to be in a quantum limbo, as long as the box is intact. The box indeed plays the most important role in this thought experiment; however, the size of the box is incorrect. The box is none other than the Calabi–Yau manifold itself! The particle's standing waves hide their energy until „measurement" (decoherence) takes place. As if these waves were frozen in time until prodded into existence, standing waves do not transfer energy and cannot be experienced until examination. Decoherence brings forth the next time moment and uncovers the (stationary) action, the energy difference between two standing-wave functions.

In everyday life, we seem to be surrounded by objects all the time. For example, we know that the objective existence of the Moon is independent of our visual experience. Material existence, however, depends on the exuberant activity of the microworld. The Moon and everything else constantly reformulates itself in the violent, incessant interactions of the quantum world due to the Pauli exclusion principle. The net result of these activities is constant aging, which makes growing old an inherent and inalienable part of existence.

EINSTEIN–PODOLSKY–ROSEN PARADOX The Bell theorem states that faster-than-light communication would be necessary to connect distant entangled particles. However, the EPR paradox expects a deeper explanation for instant communication. Entangled particles are part of one common wave function. Without the burden of gravity or time, quantum waves can spread across the whole universe. However, this quality cannot be used to send instant messages to distant galaxies. Because the particle is insulated from the outside, its quantum state is hidden until the time of the measurement, which reveals its information. As a consequence, during quantum processes, unknown information is transmitted and manipulated. But unknown information has no „informational" value.

ENTANGLEMENT BETWEEN DISTANT SYSTEMS   Entanglement between distant systems, such as between photons, has been verified in many increasingly complex experiments. For example, in 2012 Israeli scientists produced entanglement between photons that never coexisted. By entanglement swapping they entangled one member of each photon pair (Megidish et al., 2012). Because the two original photon pairs were separated in time and the second pair of photons was created only after detection of the first pair, temporal entanglement was achieved. The information embedded in the particles remained unchanged until decoherence, which is the particle's next time moment.

SLIT EXPERIMENT, INTERFERENCE   The slit experiment involves a photon (or any other particle) that passes through one or several slits on its way to a screen. A single photon is found to pass through several slits at the same time and thus interfere with itself! If we accept as fact that quantum waves exist as time and space independent waves, we can understand that quantum waves are capable of spreading out as wide as necessary and that it is energetically convenient to pass through the slits to interfere. However, placing a detector at one of the slits collapses the wave function of the whole particle, ushering in the next time moment and canceling the chance of interference.

TWO

THE QUANTUM OPERATION
OF THE MIND

Abstract

In the material world, decoherence (i.e., the collapse of the wave function of elementary particles) produces measurable changes in physical qualities, such as speed or position. However, features of the behavior of elementary particles have also been exposed in numerous mental phenomena. The particle-like and even quantum behavior is already being exploited in fields as diverse as search-engine optimization (Grover, 1997), psychology, economy, and even social sciences (Pothos & Busemeyer, 2013). This section presents evidence that the mind not only exhibits quantum characteristics in its function but it is truly an elementary fermion that interacts via elementary forces embodied as emotions. The energy requirement for mental function is inexplicably large; it is even termed the brain's „dark energy." Beyond sensory and motoric operation, the mind is primarily a temporal compass. Stimuli create temporal deficiency or excess, manifested as shifts in brain frequencies. Between past and future, the mind constantly reorients itself to overcome disturbances imposed on it by the environment. Interaction changes the temporal balance of the mind. Temporal excess increases the degrees of freedom, an empowering, energetic feeling; whereas temporal deficiency (**stress**) reduces the degrees of freedom and can remain part of the mental makeup over the long term. The essential identity of elementary fermions (matter and mind) has been obscured by the fact that material systems operate over spatial coordinates, whereas mental (emotional) changes demarcate a temporal landscape. Biological systems operate with a clock, which dictates nutritional needs and gives them an expiration date. Matter takes shape in space, ecosystems and societies form over time. Although studying particles of matter is very difficult due to their size and energy levels, examining the brain poses the opposite problem: the myriad possible ways to study it produce divergent, often confusing conclusions. However, the analog structure and operation of brain and matter holds great promise. With suitable corrections, understanding can be transferred between the fields of neurology and theoretical physics for the mutual benefit of both. This chapter is divided into five sections. The first section introduces the **energy-neutral unit** of the mind. The second section details the mechanism of mental interaction and the physics of the mind. The third section is about elementary particles of emotion. The fourth section is conclusions, and notes forms the final, fifth section.

"The universe is built on a plan, the profound symmetry of which is somehow present in the inner structure of our intellect."

–Paul Valéry

Introduction

The simple act of opening or closing your eyes changes your brain. Complex electromagnetic flows and oscillating rhythms conspire to make the mind much more than simply the cortex, the amygdala, and the other structures that constitute the brain. Immense energy consumption cannot be accounted for simply by the maintenance of the electric potential of neuronal cells and the management of their synoptic activity. Of all the organs in the human body, the mind regulates itself and successfully organizes the whole body into a seamless orchestra. Thanks to its neuronal organization, even a worm can crawl toward food and shelter. In the human brain, the sensory stimulus of sight increases oscillation frequencies. The oscillations in the limbic system project information about the environment to the cortex and back. For over a century, the electromagnetic activity of the brain has been measured by placing electrodes over the scalp, and more recently science has learned that external magnetic and electric fields can change brain activity.

The behavior of elementary particles is described by the nonintuitive rules of quantum mechanics. In contrast to classical systems, where measurement merely observes a preexisting quality, quantum measurement actively changes some property of the system being measured.

Probabilistic assessment is often strongly context and order dependent, and individual states can form entangled, composite systems. Remarkably, the same principles appear apply to the mind as well. The nonintuitive and multifarious nature of mental operations has been discussed by philosophers and sages over the millennia of human civilization. The ideas of quantum theory have been utilized in psychology for nearly 100 years. More recently, in over seventy national surveys quantum probability have been shown to be a superior predictor of human judgement. The laws of quantum mechanics turn up repeatedly in mental phenomena (Pothos & Busemeyer, 2009), and the organizational intricacy of the brain defies comprehension (Brembs, 2011). The humble suggestion that matter fermions and the mind have identical structures and identical operation is a first tentative step toward opening the book of human motivation and behavior.

Traditionally, a sharp divide has existed between studying the mind and the brain. The brain has been object of neuroscience and psychiatry, whereas philosophy, religion, and psychology focused on different manifestations of mental experiences. At the dawn of the twenty-first century, the time has come to consider the mind as a physical entity. The pages that follow argue that the brain's electromagnetic potential forms a temporal compass. Stimuli unbalance successive layers of self-regulatory processes of the mind, which nevertheless always recovers its neutral state. This energy neutrality means discrete energy processing, which turns the mind into a quantum system. Such homeostatic regulation is possible because the mind identifies itself with the body (Guterstam, 2015). If the brains of a theater audience could be scanned during a performance, nearly identical cortical activation patterns would be found in each brain. The mental activation patterns of the audience would rotate in unison, moving over the temporal landscape of the performance according to a common reaction pattern (Dmochowski et al., 2014). The details of how the brain produces such an emotional rollercoaster ride form the topic of this chapter.

Appearance of the Nervous System

Unlike plants, animals obtain food by active processes. Behind it all is the great organizer: the nervous system. Complex organs have been shown to have evolved alongside and interdependent of the nervous system.

Nerve cells are organized in two parallel columns in bilaterians (animals with left and right sides that are approximate mirrors of each other, which include all but the most rudimentary animals). Bilateral symmetry simplifies sensory processing and muscle coordination. In bilaterians, electrically coupled nerve cells are structured in dual columns and form an electric field, which is modulated according to sensory and entropic conditions. The representation of a body part in the brain is highly dependent on the body part's importance for interaction with the environment. Thus, the nervous system forms a central homeostatic regulation as it governs the body in concert with the environment. Sensory organs and the mouth are situated near the anterior end of the body (the head). Sensory information of the environmental conditions permits coordinated movement in the service of food acquisition.

The evolution of the limbic brain made fast and precise sensory responses possible, greatly improving survival. In sharks for example, limbic structures form a linear organization and lead to a fairly predictable reaction to environmental stimuli. The evaginated pallium metamorphoses into cortical structures: first the simple, three-layered allocortex and, later in evolution, the six-layered isocortex. With the emergence of the cerebrum, the linear regulation of the brain gives way to nonlinear complexity, cognition, sentience, and even intellect. Shifting energy balances activate and form neuronal connections in the cortex in a frequency-dependent manner. Repeated stimuli will never produce the same brain frequencies or mental states. In this way, the cerebral cortex allows the accumulation of emotional history (experience), which in turn produces complex, unpredictable behavior. The following section examines the connection between specific mental-energy balances and mental functioning.

Summary

The nervous system in animals provides great organizational complexity. Response to stimuli is simple and predictable in animals with a limbic brain, but the evolution of the cortex introduced nonlinear behavior. Repeated activation involves a different neuronal landscape, allowing the accumulation of experience, which results in complex and unpredictable behavior.

The multidimensional torus: Basic unit of intelligent life

The social cohesion and learning ability of mammals and birds puzzles us (McNally et al., 2012). These highly intelligent animals dominate every living environment. Birds rule the sky; in the seas the cetaceans form a complex web of social life, and on land the mammals exhibit impressive memory, communication, and organization. The cerebrum introduces a frequency-dependent temporal regulation and forms emotions, which are a sophisticated homeostatic regulation. Emotions allow these animals to be warm blooded, form the mysterious inner world of consciousness, and develop complex social life.

The dynamics of the brain changes dramatically with the evolution of the cortex. It is generally accepted that the brain's energy level is constant over time, even during sleep. For the global metabolism to remain constant, an energy decrease in one part of the brain should correspond to an energy increase in another part of the brain (Raichle & Snyder, 2007; Mantini & Vanduffel, 2012). Thus, science has consistently found that, while performing tasks, **default-mode-network** (DMN) activity decreases. Furthermore, such constant metabolic activity (a difficult regulatory feat) seems to characterize many—probably all—mammals. As the reader will recall from Chapter 1, such a constant energy level is an essential feature of elementary fermions.

Summary

The regulatory complexity of the cerebrum permits birds and mammals to form emotions, which allows the accumulation of experience and learning. The brain's energy level remains constant over time; this energy neutrality is typical of elementary fermions.

Unity of the mind

Elementary particles are the smallest units of energy and cannot be subdivided. Descartes, Kant, and others predicted that unity has to be an essential feature of the mind. The body's representation in the brain allows a feeling of oneness with the body (Guterstam, 2015). Ideas and thoughts form a highly fluid, malleable mental background over which interaction with the outside world becomes possible. The mind is a cacophonous sensory kaleidoscope, peppered with transient ideas and possibilities that distill into a single decision or understanding. The sensory „forest" coalescences a single, unified experience. For example, at any given time, only one view of the Necker cube can be valid (**Figure 2.1**). Indeed, although we can contemplate many possibilities; once we decide on a problem all other options cease to exist.

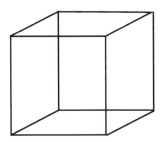

Figure 2.1. Necker cube The Necker cube is a wire frame cube with no depth cues. It is impossible to hold two views in the mind, so the mind will settle on one of the two possibilities.

How does unified experience emerge in the mind? Neuronal activation can occur either by temporal coding (the synchrony of oscillations) or rate coding (the increasing frequency of oscillations). Temporal synchrony can powerfully generate postsynaptic oscillations. For example, simultaneous activations of a target ne-uron by two presynaptic neurons results in exponentially greater activation than the algebraic sum of the two initial activations. In contrast, repeated activations by one presynaptic neuron leads to post synaptic depression. In this way, sporadic, disjoined activations die out, and wide spread, congruent activities converge toward a unified, correlated mental picture (Ainsworth et al., 2012). Transitive neuronal assemblies are capable of still-higher-level synchronization by forming a nonlinear, highly dynamic, and multilevel organizational pyramid, which leads toward a unified yet highly abstract experience.

As early as 1957, the powerful inner drive to maintain cognitive consonance was recognized by Leon Festinger. His cognitive dissonance theory states that incongruent belief or behavior forces mental change to avoid the frustration of cognitive or emotional discrepancy. Even core beliefs can be sacrificed to maintain or restore mental congruency. The constancy of self becomes particularly apparent when changes, even dramatic ones, affect the body or the brain.

The Calabi–Yau torus exhibits holographic organization. The manifold of matter fermions is a holographic (temporal) record of spatial changes. The cortex formulates a „temporal horizon," which becomes the memories and accumulated experience of a constantly changing cortical projection. Of the billions of photons hitting the retina and the millions projected to the optic nerve, only a few thousand bits of information or even fewer produce the conscious perception of the moment. Therefore, consciousness forms on a momentarily changing and highly subjective (holographic) mental landscape, unknowable, with the power to surprise even the self. Indeed, we often cannot know or cannot make up our own mind. The momentary projection of the temporal manifold (**subconscious**) depends on both the viewer and the self. The holographic self is more submersed in memories and the neural potential of the cortex than an iceberg is in water.

Summary

As the smallest units of energy, elementary particles cannot be divided. The mind also forms a unified sense of self over time, despite substantial changes to the body and the constantly changing sensory stimuli it receives from the environment. The history of material systems is recorded in the information memory of the Calabi–Yau manifold. The manifold of the mind is the neuronal organization of the cortex, where the interactive history of the organism accumulates.

The brain as camera obscura

The camera obscura is a box with a pinhole through which a small, inverted image is projected onto its inside wall. Surprisingly, the brain processes sensory stimulus in a similar fashion. Both sensory stimuli and motor innervation are reversed in the brain. We see images upside down, and our brain learns to invert the image in infancy. Other sensory input is also processed upside down in the cortex. Sensory stimuli from the head and face correspond to the lowest part of the parietal lobe. Similarly, the motor cortex at the back of the frontal lobe innervates the body upside down. In the primary motor cortex, the muscles of the head and face are innervated at the bottom, and the legs are innervated at the top. For any perception we wish to examine, the cortex behaves as the back wall of a camera obscura, on which a smaller, inverse image of the sensory reality is projected. For instance, on their way to the cortex, sensory and motor nerves switch sides in the brainstem or the spinal cord, which thus functions like the camera's pinhole. The cerebellum does not cross, but its connections toward the cortex do. Examining the energetic landscape of the subconscious manifold, the advantage of the above arrangement becomes clear: The innervation of the left and right of the head, face, and hands occurs at opposite sides of the cortical energy field, ensuring maximal sensitivity for the hands and face. Nerves for the sensory and muscle coordination of the legs are found at a small energetic distance from each other on the top of the head, where the continuous field mutes differences, resulting in better balance. This arrangement increases sensitivity between the two sides of the face and the two hands and enables finer manipulation; yet interhemispheric fibers maintain synchronization between homologous brain regions (Buzsaki et al., 2013). However, the field at the conjoined cortical areas at the top of the head improves coordination and synchronization of extremities in spite of their significant physical distance from the center of the body. Notable exception is olfactory bulb, which is located in the limbic brain. Thus, the sense of smell lacks spatial dimension and has a linear nature. But because of its limbic origin, smell is more visceral, affecting us on an elementary, instinctive level.

The reversed innervation of the muscles serves another important role. Contralateral force generates an inward vector, leading to structural stability. In engineering, this is such a fundamental concept that it is used in countless applications, from simple scissors to robotics, cranes, and big loader machinery. This makes bilaterians successful and dominant in evolutionary history. Even the simplest bilaterians demonstrate bodily cohesion because the central nerves (where the field is most stable) innervate the sides furthest from the center of the body.

Summary

The cortex behaves as the screen of the camera obscura with respect to sensory processing and motor coordination. The body is represented upside down in the cortex to provide better sensitivity for the face and hands, which are processed as polar opposites at opposing sides of the cortex. The extremities of the legs are processed at the top of the head, where the common field mutes differences allowing better coordination. An exception is the sense of smell, which, being processed in the limbic brain, forms a linear organization.

Brainwaves

In response to stimuli, neurons in the brain spontaneously form complex oscillating patterns, generating electric and magnetic fields that can be measured. The first human electroencephalogram (EEG) recording has transpired in 1924 by Hans Berger. The function of oscillations in mental operation is well established, but their exact role and relationship has been a subject of intense debate and study. The complex network of electrical activity in the brain is energy intensive; the brain uses ten times more energy per unit mass than any other tissue, using over twenty percent of the body's energy intake. But neuronal processing of stimuli accounts for only a portion (estimated to be 20% to 40%) of the brain's large requirements in energy, the vast majority of neuronal operation is poorly unaccounted for or unrelated to the task at hand. However, careful and ingenious studies over the past two decades have shown the crucial, interrelated relationship between the brain's function and its frequencies. In the working brain, long-range connections, supported by subcortical neurotransmitters can easily shift energy balances (i.e., information flow) to produce coherent, flawless operation. Such complex brain activity is an essential requirement of consciousness.

Brain waves can be separated into well-defined ranges of frequency. Delta waves are less than 4 Hz, and theta waves are between 4 and 7 Hz. These two types of waves are typical in children. In adults, delta and theta waves occur during deeply relaxed, meditative, creative, or minimal-energy (minimally conscious) states such as some phases of sleep, anesthesia, and coma. Alpha waves (8–12 Hz) typically occur when the eyes are closed, or when relaxing. Beta waves range from 12 to 30 Hz and are associated with concentrated, worried, or active states. The highest-frequency (39–100 Hz) gamma waves occur during movement, such as complex motor activity, attentiveness, and detailed sensory perception, but can also correspond to excited, conscious focus, or nervous emotional states. Gamma frequencies are often interwoven with lower-frequency oscillations; information transfer occurs by entrainment of gamma-band synchronization (Engel and Silva, 2012). The importance of brain oscillations in decision making and consciousness is unequivocal. As discussed in the „Heuristic foundation of hypothesis" in Chapter 1, the energy direction in elementary particles is frequency dependent. In the brain, the direction of information (energy) transfer in the limbic structures is highly dependent on frequency. Low brain frequencies intuitively increase the degrees of freedom; whereas high brain frequencies are more deterministic and therefore allow fewer degrees of freedom (Buzsaki et al., 2013). Therefore increasing frequencies contract, forming down spin, and decreasing frequencies expand, forming up spin. Similar balances in energy (contraction speeds up movement and expansion slows down movement) govern every spinning body, be it a top, a spinning ice-skater, or an electron. In the brain neocortical-limbic transfer occurs during slow theta waves (4–10 Hz), and data transfer reverses during gamma frequencies (30–130 Hz), as reported by Buzsaki (2011), among others. Because energy is proportional to frequency, the brain essentially „pays" for sensory information. The classic equation

$$E = h v \text{ (Equation 2.1)}$$

shows that energy E is a function of frequency v. But here this means even more: the environment oppresses and hijacks the mind to extract energy in exchange for information. Reading and comprehending signs or deciphering other visual and auditory stimuli occurs automatically, without conscious intent. In fact it takes conscious effort to *resist* interpreting sensory information. Once we learn to speak a language, we are enslaved into comprehension and understanding. Thus we can only recognize the musicality in intonation of languages that are completely foreign to us.

„Heuristic foundation of hypothesis" in Chapter 1 also pointed out that up spin insulates the torus, so it remains a local force. In the down spin state; MiDV is transmitted over the field as a photon. However, the orthogonal mental fermion should form identical operation over time, rather than space. Up spin (low brain frequencies) would instantaneously produce mental volume or **mental energy**, which is a wealth of time; a powerful, happy, satisfied, and harmonious mental state. Down spin of high brain frequencies (beta and especially gamma) would form temporal bosons, causing excessive (unnecessary) details, leading to the sense of a shortage of time. Mental volume would contract by forming information. The connection of enhanced brain frequencies with negative mental states has been corroborated in many studies (Bethell et al., 2012; Seo et al., 2008). Consequently, the mind polarities are formed by positive and negative emotions. Today, external electric, magnetic field or ultra sound called brain stimulation, is increasingly utilized to modulate brain waves for therapeutic or recreational purposes.

However, after treatment, Ewing and Grace (2013) found significant „rebound" effects in both power and coherence in multiple brain regions, demonstrating the steadiness and stability of brain frequencies. Since the brain's structural connectivity and, by extension, thinking and behavior reflect our mental make-up and long-term emotional history, brain oscillations show high stability and even characteristic of the individual. This might explain the hereditary aspect of brain frequencies which, independent of education, remain stable throughout life and it is one of the most heritable characteristics in mammals (Buzsaki et al., 2013; Fingelkurts & Fingelkurts, 2013).

Summary

Brain oscillations can be grouped according to their frequencies. Lower frequencies are typical of sleep and relaxed states, whereas higher frequencies are characteristic of enhanced focus or troubled conditions. High brain frequencies move information toward the cortex, whereas low brain frequencies reverse information flow. The evolutionary conservation of brain frequencies in mammals reflects their vital importance. Drawing on our understanding of material particles, lower brain frequencies should form up spin conditions, while higher frequencies should form down spin conditions. However, the greater energy requirements of many brain functions also require greater frequency oscillations. For example, the brain pays for sensory information by the energy requirement of higher frequencies. Nevertheless, as shown later, brain frequencies determine the spin direction of the mind.

Anatomy of the mental gyrocompass

Matter fermions orient against the SF and interact via fundamental forces. The mind is regulated by its emotions, which instigate changes and force actions until emotional stability and neutrality is achieved. Like up spin states of matter fermions, positive mental states are transient. Only negative emotions involve the long-term presence of temporal bosons. Emotions are the tools of mental elementary particles for automatic orientation over the TF. This is the basis of the mind's homeostatic regulation.

A gyrocompass is a compass based on a gyroscope and is used on a planet or other spinning galactic object to determine orientation. Unlike a magnetic compass, a gyrocompass is based on planetary rotation, so it points toward true north. It is oriented by gravity or by using a spinning wheel situated in a viscous medium. As the planet turns, misalignment causes tilting to minimize the potential energy, which orients the gyrocompass toward true north. Like a gyrocompass, the mind shows a cunning ability to maintain the stability of the inner world of consciousness against relentless bombardment by outside stimuli. By changing the *mental energy balance* (the neuronal landscape, manifested as the ratio of quantum-manifold energies and the TF curvature) through wave-function collapse, the mind recovers an energy-neutral state. *In this way the mind changes constantly and gradually with its environment.* The outermost layer of the temporal gyrocompass is the brainstem, which integrates the mind into the environment through its connection to the sensory organs and the body. Its gate-keeper role gives the brainstem an important role in influencing higher brain functions. Information transfer toward the cortex is regulated in the limbic brain which, through sensory and motor regulation, forms the middle layer. Cortical activation forms the third, innermost layer. This is the self which, through sensory processing, identifies itself with the body and becomes the source of self-awareness and forms the **ego** (Guterstam, 2015). This is the transient, unknowable, and magical inner world of consciousness. In highly intelligent mammals, such as primates and dolphins, the limbic brain is embedded under the umbrella of the folded cortex. This chapter analyzes this highly evolved brain, paying particular interest to the human brain. Let's now meet these three layers of the mind.

THE BRAINSTEM The pons, midbrain, and medulla are parts of the brainstem; an organ sandwiched between the body through the spiral cord and the lower limbic areas of the brain. The midbrain plays a role in evolutionarily conserved regulatory functions, such as posture, balance, breathing, sleeping, and heart rate. The midbrain forms the pinhole of the mental camera obscura. Depending on the direction of the connection, sensory and motor innervation crosses in the midbrain to the opposite side of the brain or body.

As discussed in the section „Mass" of Chapter 1, material fermions form their local SF strength; in turn, the field regulates their ability for interaction (Figure 1.7). The SF forms gravity, whereas successive events formulate the TF (temporal gravity), which regulates biological systems from plants and animals to man. In vertebrates sensory input arrives at the brainstem from the sensory system, giving it a front row seat for a quick, overall impression from the senses. Midbrain structures, such as the ventral tegmental area play an important role in motivation and automatic responses. With its extensive connections to other parts of the brain such as the amygdala, the brainstem can directly modulate attention, sensory function, and mood. This plebeian part of the brain quickly evaluates the environment for its entropic value and lends us a „gut feeling," or sixth sense. Thus, the perception of the environment is typical for the age, sex, and personal characteristic of the individual. The brainstem's fast sensitivity to the environment plays a decisive role not only in our momentary responses but also in our sensory follow up. Consequently, ongoing brain activity significantly influences or even determines perception by tuning the brainstem's sensitivity toward stimuli (Oohashi et al., 2000; McCraty & Atkinson, 2014). Activity in the sensory cortex biases subsequent sensory detection. For example, cortical activation inhibits auditory detection (by competition), whereas activity in the DMN, such as the posterior cingulate, facilitates it (Sadaghiani et al., 2009). However, the brain with elevated brain frequencies forms a hypervigilant, insecure mind. As will be shown later, relentless examination and attention to detail is used to verify mental bias. This way, mental response is more dependent on the state of the mind than on the sensory and motor attributes of the signal. I propose that, by forming a temporal weight, the mind (brainstem) seeks constant alignment (a minimal-energy configuration) to the local TF. During this process the mind shifts its mental disposition, until the temporal weight and the TF curvature of the environment is congruent. Hence, our mental makeup dictates our position in society.

Summary

The brainstem connects the limbic brain with the spinal cord and the body. Sensory and motor innervation crosses in the midbrain to the opposite side of the brain or body, depending on the direction of the connection. The brainstem also controls primitive, ancient, and evolutionarily conserved regulatory functions, including breathing, sleeping, and heart rate. Our attitude dictates the brainstem's sensitivity to stimuli, which often has greater effect on our behavior than the sensory attributes of the signal. The mind forms its own temporal weight, which determines the difficulty of life.

THE LIBMIC BRAIN   The constantly changing frequencies of the limbic brain form the second layer of the mental gyrocompass. Its modular, essential structures provide bodily functions, emotion, and conscious regulation. Some of these structures will be discussed in connection to their physiological role. The sensory organs project from the body to the limbic brain, where spatial information is transformed into temporal rhythms to be transferred to the cortex. In the mammalian brain, frequencies are complex, interconnected rhythms, with higher frequencies building on and interlaced with lower frequencies.

CINGULATE CORTEX Strategically located between the cortex and the limbic brain, the cingulate cortex has a vital role in reward, pain, emotion, error detection, and behavior (Vetere, 2011). Damage to this structure often robs patients' emotional lives, leaving them emotionless and expressionless (Njomboro, 2012). Like spokes of a wheel, axons form extensive connections between the cortex and limbic areas, as demonstrated beautifully by tractographic studies of the brain (Filler, 2010; **Figure 2.2**). Information (i.e., energy) flows between limbic and cortical modules across the cingulate cortex (gyrus), which gives it a central role in emotional regulation. As shown earlier, the flow of information is frequency dependent. The intensity of the energy flow between conscious and subconscious brain structures is proportional to the intensity and urgency of emotions (Basbaum and Fields, 1978), which are named based on their sensory context and the cortical-cerebral memory library.

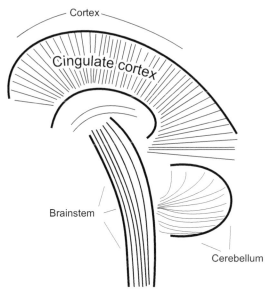

Figure 2.2. Neuronal connections in the brain Tractographic diffusion tensor imaging measures water diffusion within living tissue. In the brain, water diffusion follows precisely the direction of the axons and delineates them. Thick lines outline structural elements of the brain, whereas thin lines indicate bidirectional water movement inside the axons. Radial lines between the limbic brain and the cortex indicate the cingulate cortex.

Summary

Limbic structures play a role in sensory, motoric, and emotional regulation. The information and energy flow in the limbic brain depends on the brain frequencies. The cingulate cortex forms a ring-like structure between the limbic brain and the cortex, creating a strategic location to register energy flow to and from the cortical region. By cataloguing brain frequencies, it has a central role in emotional regulation, reward, error detection, and behavior.

THE CORTEX   The cortex evolved from the evaginated pallium, where neuronal assemblies are activated in a frequency dependent manner, allowing the activity to be repeated later in an automated fashion. The accumulation of records of neuronal activation (the development and organization of synapses) becomes the basis of mental operations and learning, which opens the way to increasingly complicated behavior. For example, in puppies, automatic muscle twitches trigger the innervation of muscles. In turn, neuronal paths form memories of the resulting limb movements. The tentative initial nerve activities that are initially strongly shaped by the environment gradually develop into automatic neuronal activation patterns, leading to concerted and automatic muscle innervation, which is movement. The ability to perfect itself through interaction with the outside world is a remarkable characteristic of the cortical brain. The considerable plasticity of the mind is a great recent discovery of neuroscience. Pathway stimulation induces new neuronal connections, which allows learning to occur in as little as a few hours. The recovery of a greatly damaged brain, even decades after the original infliction, is a remarkable possibility (Kantak et al., 2012). As healthy brain areas take over the functions of the damaged parts, the remaining brain tissue (sometimes of greatly reduced volume) can support full, or nearly full, mental functioning. The astonishingly great plasticity of the mind raises the possibility that, in spite of brain damage, the energy unity of the mind, or the mental self, remains intact.

I propose that electromagnetic oscillations form in the brain as sound vibrations form in a drum or flute. Just like in wind instruments, the higher notes are produced by reducing the vibrating volume. In the brain, higher frequencies are produced by activating a smaller number or a smaller volume of the brain's modular structures. Just as a particular instrument can be recognized by its individual sound, brain oscillations (particularly in the lower-frequency range, which are produced by greater brain volume) are presumably characteristic of the topographical shape and volume of the cortex. Thus, the frequencies of the brain can be considered to be universal notes for functioning brains. Nevertheless, the characteristic range of brain frequencies changes over our lifetime as a function of gender or individual history. For example, the lowest brain frequencies gradually disappear over the years, a process more pronounced in male than in female mammals.

The connection between intellectual abilities and the complexity, convolution, and overall size of the neocortex is well demonstrated (Deaner et al., 2007). In animals, a rough positive correlation exists between brain size and body size. However, the cerebrum's modular structure in birds is inefficient for growth, which effectively curtails the size of their brains. Nevertheless, the small but efficient brain of birds combined with their agile bodies turns birds into the acrobats of the sky. Neurons organize in layers in the mammalian cortex; a structure that can easily expand with growing brain size. During mammalian evolution, this organizational flexibility permitted a big expansion of body size and brain size. Although four orders of magnitude separate the brain size of shrews from that of whales, the typical brain frequencies have remained constant. The only frequency band that shows considerable difference is the theta band, which is lower in humans (1-4Hz) than in small animals such as rodents (6–10 Hz), as noted by Buzsaki and colleagues (2013). Although axons with large conduction velocity facilitate information transmission, the remarkable evolutionary stability of brain frequencies in spite of enormous differences in body size, brain size, and intellectual abilities among mammals is not yet understood. However, neuronal activation on the cortical surface absorbs and extinguishes brain oscillations. Therefore standing waves form between the limbic area (where the brain frequencies are generated) and the cortical surface. The formation of standing waves requires exact volume (or constant surface-volume ratio). However, brain volume increases as the radius cubed, whereas the cortical surface increases as the radius squared. As with wind instruments, in which the length is changed to modulate the sound, the evolutionary stability and universal regulatory code of brain frequencies is ensured by the increase of the cortical surface area (and square root of thickness) as the third power also. Recently the brain mass and cortical surface area was found not to form a linear relationship (Mota et al., 2015). However, accounting for the thick cortical mass of cetaceans (a substantial addition to limbic mass) and smaller brains (which remain non-convoluted up to 400 square mm surface area), then the linearity of the cortical surface area to limbic volume (mass) might be uncovered. The packaging of a disproportionally increasing cortical surface led to convolution, which is the formation of gyri and sulci, which inadvertently produced a more robust cortex.

Indeed, the most convoluted cortices can be found in animals with the biggest brains. For example, the cortices of elephants and cetaceans are more convoluted and larger than those of humans. The folded cortical structure inadvertently leads to greater mental capacity, information capacity, memory potential, and learning.

About a century ago, Spearman introduced the idea of mental energy to explain the positive correlation found in various intelligence tests, and his ideas remain relevant today. Like Spearman, I consider **mental energy** to be an intellectual ability. It corresponds to degrees of freedom in physics; it is the ability to process and control new information with openness, freshness, and trust, which enables mental freedom. Later, I show that mental energy is a subconscious (cortical) ability. The greater cortical insulation also leads to emotional stability and sophistication. Thanks to these advantages, large mammals with convoluted cortices form close-knit, stable societies based on compassion and kinship. The constantly changing brain activity allows the accumulation of experience by the immensely complex neuronal connections of the cortex. To face new challenges, old associations can be reconnected in a novel way, until the pursuit of the new becomes of elementary mental importance.

Summary

The cerebral brain has a remarkable ability to perfect itself. Interaction with the outside world accumulates experience through building and modifying neuronal connections. The cerebral bird brain has a modular structure, allowing learning and intellect in a small package. The mammalian brain evolved a laminar cortex, which allows an almost limitless flexibility in size, yet the brain's frequencies have remained constant over evolution from shrews to whales. The brain's volume increases as the radius cubed, whereas the cortical surface area increases as the radius squared, leading to a smaller brain frequencies with increasing brain size. The brain keeps the ratio constant by increasing cortical thickness and/or the cortical surface area, which fits around the limbic structures by convolution, called corticalization. By lending emotional stability and intellectual ability (mental energy), the cortical thickness inadvertently benefits the organism, leading to a confidence of openness and trust, which is the ability to process complex social information. Higher brain frequencies are produced by fewer neural modules, whereas the smaller frequencies involve greater brain volume. Lower brain frequencies gradually diminish in the later years of life.

The mind as a temporal fermion

The great discovery of the past three decades is that the traditional idea of intellect does not fully explain the differences in personal, professional, and economic success. Social and emotional intelligence often show similar or better correlation with achievement (Barbey et al., 2014). In scientific literature, from psychology to neurology, the relationship between emotion and cognition to achievement is increasingly being recognized. But principles of classical logic also fail when consciousness is examined. Human decision making can be best described by quantum probability, allowing the successful use of quantum theory in, for example, search-engine optimization (Pothos & Busemeier, 2013).

According to string theory, energy vibrations take shape as matter, but in the material brain life is an affair of energy. A stimulus is processed as the energy of oscillations, allowing the brain to formulate its operation based on energy. Highly structured frequencies reveal nonlinear structural complexity, which nevertheless formulate an energy-neutral state, the DMN, corresponding to the energy-neutral structure of the Calabi–Yau torus: an elementary fermion. Material fermions align themselves to the SF, whereas the mind relies on time. Environmental changes formulate the TF, which constantly modulates brain frequencies via the sensory organs. The fluidity of the inner mental world is punctuated by interactions, which section our mental life into discrete „states of feelings and beliefs," which form a mental progression and gives life an irreversible directionality. Just as for matter, life hinges on interactions. Decoherence reformulates standing waves by increasing or decreasing mental energy (mental volume). Elementary particles interact by elementary forces. The elementary forces of the mind are emotions, which are inextricable phenomena of life whose intensity can only change through interaction. They may disappear from conscious awareness, but the boson of negative emotions manipulates the mental state from the background by creating stress, which corrupts mental abilities (recall that up spin states and positive emotions are transient).

Despite matter and mental fermions having identical energetic structure, several important differences separate them. Perhaps the most important difference between matter and emotional fermions is their size, which effectively determines their energy level. The diminutive matter fermions give rise to enormous frequencies, which produce an impressive punch. The far larger mind forms much lower frequencies and energy levels that are many orders of magnitude smaller (we cannot bend a spoon with our thoughts).

Material fermions are insulated from gravity, whereas mental quantum waves, shielded from temporal gravity, exist unlimited in time, the past or the future. Thus, material quantum waves enjoy spatial freedom, whereas mental quantum waves revel in temporal freedom. This is the reason that emotions are endowed with a sense of permanence. Pain or joy feels as if it would exist forever, but when emotions depart, their experience evaporates, as if they never existed. This fact has a great role in motivation. By feeling permanent, emotions propel actions, but their fleeting nature allows us to find new strength even after immense pain and suffering. The temporal quantum waves of matter interact with space to produce time and the brain frequencies interact with time to formulate mental volume. For this reason, decoherence gives spatial expanse to material systems, but mental scope, expanse, and understanding are temporal. The difference between the manifestations of matter and mind effectively has hidden the symmetry between the two systems.

Summary

Over the past three decades emotional and social abilities have begun to be recognized as the crucial ingredients of intelligence. Material fermions and the mind have identical structure and operation, but their size difference determines an immense deviation in energy level. While matter takes shape in space, the mind changes over time. Therefore emotions are endowed with a sense of permanence, but this is only an illusion. Emotions motivate and propel actions by their sense of permanence, but their fleeting nature permits us to renew our strength and to go on with life.

THE TEMPORAL FIELD   The tiny compass of the electron orients not only along the electric field but also along the SF. Living organisms are affected by both spatial and temporal fields (entropy). Plants grow and animals maintain their balance against the SF (gravity). In mammals, the vestibular system serves this purpose. The neural system, however, is a tiny temporal compass that aligns with the TF, between negative and positive time. **Macrodimensional time** (MaDT) and **microdimensional time** (MiDT) form the TF. With the evolution of the brain, the brainstem came to be the entropic sensory organ that aligns with the TF. Because the ratio of compactification to expansion defines local gravity, so the ratio of positive to negative time defines **temporal gravity**. Temporal gravity is formed by positive TF curvature;

it is characterized by the presence of MiDT, which overwhelms us with details, making problems—even the easiest ones—difficult. Although large temporal gravity is constricting, its absence produces carelessness and superficiality.

An example of entropic regulation in the brainstem is the periaqueductal gray (PAG). The brainstem assists the cortex to evaluate the body and its physical status. Electrically coupled cells in the reticular activating system (RAS) provide feedback to the motor cortex about the conditions (entropic value) of movement; mistakes in movement result in arousal and correction. The PAG also serves an essential role in pain regulation. Pain is an important entropic signal of the positive-curving TF that is registered as increasing frequencies in the cingulate cortex. During analgesia, synthetic opioid drugs inhibit signals in the pain pathway in PAG from reaching the thalamus and the anterior cingulate. In a study on rats, Simon-Thomas and colleagues (2011) demonstrated the PAG's regulatory role by artificially producing an entropic seesaw (negative or positive time). Stimulation of the dorsal and lateral aspects of the PAG can provoke defensive responses, which are characterized by freezing, running, jumping, tachycardia, and increased blood pressure and muscle tone. This positive time (MiDT) galvanizes the fight against invaders and can turn into aggression and, at its extreme, petrified immobility (asymptotic inhibition of activity). The changes demonstrate the effects of increasing TF strength. In contrast, stimulation of the caudal ventrolateral PAG coincides with negative time (i.e., MaDT). The wealth of time leads to an immobile and relaxed posture known as quiescence. These changes nicely demonstrate the inescapable self-assured mental confidence that come from the sense of the richness of time. Correspondingly, lesions of the caudal ventrolateral PAG were found to greatly diminish conditioned freezing, whereas lesions of the dorsal aspect reduced innate defensive behavior, effectively „taming" the animal (leading to the trust of a white hole state!). These results reveal the fast, automatic (nonconscious) regulation of the mental state by environmental entropy, which is exerted through the PAG. Just as spatial curvature (gravity) controls matter, the temporal balance inevitably regulates the mind.

The source of matter is decoherence. Without it, caresses could not be felt, and the light of the moon could not be seen. Decoherence is also the source of life: without interactions, we die. To satisfy the physical needs of the body and the emotional needs of the mind, decoherence is essential. In white holes, decoherence produces enormous energy transfer, which slows down within Euclidean environments and comes to a crawl near black holes. The same energy picture is true for the mind. Infancy is an emotional white hole state, characterized by low brain frequencies. The immense energy changes correspond to intense emotional states. Boundless joy and excruciating pain make the emotional life of children intense. The energy change of interaction progressively declines in old age and mutes emotions, so the difference between joy and pain decreases (Figure 1.13, black hole state; in the case of the mind the x axis should be marked TF, rather than SF).

Summary

The gravitational field not only regulates nonliving matter but also living organisms. Animals maintain their balance against the gravitational field. Environmental changes form the TF and determine a temporal gravity, which is countered by the brainstem as the mind sets its own sensitivity for environmental entropy. Areas of white holes are characterized by enormous energy changes. The mind forms immense emotional states in infancy, which is our white hole state. In older matter the smaller energetic changes slow the inner clock. In the same way, frequency differences are smaller in older people, resulting in less pronounced emotional swings (joy or pain).

Mechanism of Mental Operation

The SF determines the behavior of matter. Material interaction forms the TF (the sequence of events), which in turn governs living organisms. To ensure survival, even the most primitive animals orient themselves along the TF (between the new, formed by for example food, and the safety of the familiar). The cerebral brain forms an energy-neutral unit, called the mind. Brain activity is a function of interaction, through stimuli, with the outside world. With the emergence of the mind, the preservation of the self (ego) becomes the major motivator which, in extreme cases, can become incongruent and even contrary to the survival of the individual and the species. The primary sensory cortex gives rise to a complex pattern of electromagnetic flows and fields, which intertwine to stimulate and modulate neural activity in an automated fashion. Since low- and high-frequency bands determine opposing energy-information flow, they can be considered as opposite energetic poles of the brain's operational regulation. The high energy need of enhanced brain frequencies curtails the volume of vibrating brain tissue, whereas the energy transmission capacity disappears during the lowest frequencies. The inverse relationship of energy-information transmission leads to a power-law distribution between frequency bands (Penttonen & Buzsaki, 2003). Phase synchronization can form common oscillatory fields of neuronal ensembles. These clusters have sparse connectivity between them, which prevents the activity from spreading into the whole network, and keeping a topographical and temporal semiperiodicity. Since sporadic modules disintegrate, cross-frequency synchrony supports associative representations (Fingelkurts & Fingelkurts, 2015).

   Some examples of the automatic regulatory potential of the electromagnetic flows of the brain are listed below. These effects clearly demonstrate that the brain's electromagnetic flows are not the byproduct of brain operation but constitute an inherent, essential regulatory mechanism: the source of mental operation.

(1) In spite of the tens-of-thousands-fold increase in brain size; the frequencies of neural communication have remained constant, which gives a clue to the syntactical coding of energy and information by neuronal messaging.

(2) The parallel organization of pyramidal cells aided by inhibitory interneurons can induce large local field potentials, which can be detected outside the brain by EEG or magnetic electroencephalograms,

or by electrodes placed inside the brain.

(3) The brain's own frequencies form electric (and magnetic) fields that influence the activity of neural assemblies, modulating the phase, amplitude, or both (Buzsaki & Wang, 2012). Endogenous field effects (local field potentials) can have measurable physiological effects on brain activity in vivo (Radman et al., 2007; Weiss and Faber, 2010). For example, neurons fire coherently even in the presence of weak (<0.5 V/m), uniform AC fields (Makowiecki, 2014). Anatomical connectivity between distant regions of the brain modulates electromagnetic potentials, which influence axonal propagation.

(4) There is a causal relationship between EEG oscillations and cognition (Fingelkurts & Fingelkurts, 2015). External magnetic and electric fields can transiently modulate behavior by altering neural activity. Transcranial stimulation has real, measurable physiological effects. Animal behavior can be reliably directed by implanted electrodes.

(5) EEG is characterized by constant-amplitude, frequency, and phase intervals alternating with abrupt jumps, which indicates the existence of a discrete transfer (in quanta) of energy and information (Fingelkurts & Fingelkurts, 2015).

(6) Decisions or specific actions can be successfully predicted based on the firing patterns of neuronal populations. For example, brain activity prior to a stimulus can predict the outcome of the stimulus (Murakami, et al., 2014), and modulating brain activity can modify behavior (see section „Free Will," below). The close connection between neuronal activation and imminent actions is a strong indication that the *brain's appropriate electromagnetic potentials, manifested as interconnected oscillations, produce those actions and decisions!*

(7) The brain always recovers its energy-neutral DMN, which is present in all mammals (Mantini & Vanduffel, 2012). The brain's DMN stereotypically deactivates during sensorimotor or cognitive tasks (Sadaghiani et al., 2009).

(8) Stimulus is built on the brain's instantaneous electromagnetic flows. For this reason, even identical stimuli would induce different electromagnetic flows in the brain.

(9) Mental spin states are well demonstrated to affect degrees of freedom in diametrically opposed ways, see „Temporal field: Entropic regulation by periaqueductal gray (in box)." The low and high frequencies should calibrate to achieve energetic neutrality over time. Therefore competition between frequencies should negatively influence brain function. For example; alpha waves (low frequencies) have been shown to inhibit ongoing sensory brain activity (Mathewson et al., 2012) and the

increasing energy requirements of sensory, cognitive, and analytic focus inhibits syntactic congruency (Fingelkurts & Fingelkurts, 2015).

(10) Place-cell sequences of the hippocampus activated during movement are replayed forward before running and backward following movement. Since population bursts spread by excitatory activity, the only way to reverse the discharge sequence is to reverse the voltage gradient (Buzsáki & Silva, 2012).

(11) Unconscious stimulus can influence brain activity and even produce physiological results (Oei, 2012).

Summary

In mammals, the energy-insulated limbic system forms the mind and motivates self-preservation. The energy required for high frequencies limits the vibrating brain volume, whereas the information-carrying capacity of low frequencies disappears. As a result, the brain frequencies show a power-law distribution. The electromagnetic flows in the brain give rise to constantly changing energy balances, the source of an automatic regulation.

THE ANATOMY OF DECOHERENCE The mind, a temporal Calabi–Yau torus, forms energy-neutral standing waves over time. Thus, brain oscillations can be viewed as a spring that moves energy (or information) in the form of electromagnetic current between the cortex and the limbic brain, always restoring the energy-neutral equilibrium position, called the DMN. The innate drive toward energy neutrality leads to a subtle regulation by the continuous and pervasive electromagnetic flows of the intact brain, giving rise to inexplicable and mysterious mental operation. The mental energy balance remains congruent with the TF curvature, which turns the mind into a temporal gyrocompass. Thus changes in TF lead to temporal energy vacuum (MaDT or MiDT) in the mind. A positive stimulus (negative TF curvature) leads to MaDT, whereas negative stimulus (positive TF curvature) results in MiDT. As already mentioned, MaDT corresponds to positive emotions, whereas MiDT parallels negative emotions. Emotions force actions that restore the DMN on a different manifold-quantum energy ratio (different mental energy). The mechanism of switching between standing waves was discussed in the section „The Heuristic foundation of hypothesis" (see Chapter 1). In this way, the mind forms standing waves that are true to the local field.

The mechanism of decoherence in the brain is detailed below.

(1) Positive stimulus forms low brain frequencies, which flows information away from the cortex (but flows energy toward the cortex), toward the limbic areas, and out to the environment. This increases the degrees of freedom through long-term depression of synaptic strength, for example (Dudek and Bear, 1992). The MaDT of up spin decoherence insulates the torus, thus the discharged energy is used up at the moment of its creation, which promptly restores the energy-neutral state (i.e., the DMN). This is the reason, happiness cannot be contained. The positive emotional states, such as joy, kindness, relaxing, playing, embracing, and even generosity project emotional energy into the environment, thereby recovering the energy-neutral state of the mind! This outward energy flow is manifested as an outward focus, comfort, and trust. Energy can also be expended as physical movement, such as laughing or jumping for joy. The temporal spaciousness of lesser TF curvature enhances mental energy by instilling confidence and wisdom.

(2) Sensory information does not come cheap. The brain „pays" for sensory stimuli through the energy required by enhanced brain frequencies. The limbic system channels incoming stimuli (information), as fast oscillations, to the sensory cortical surface, building an electric potential difference between the limbic and cortical areas. Brain frequencies are electromagnetic flows that are transferred onto the cortical surface, where they form electric currents that accentuate or subdue each other through field effects. The brain's highly fluid neural organization allows fast, although not instantaneous, rebalancing of electromagnetic gradients based on charge conservation. From the sensory cortex the oscillations spread on the cortical surface, propagating toward the frontal associative regions. The energy requirement of neuronal activation extinguishes the information flow. Due to the newly formed electric potential difference between the cortex and the limbic region, slow oscillations reverse information flow from the frontal toward occipital direction, and back toward the limbic region in order to recover the DMN. The temporal standing wave formed by sensory transmission toward the sensory cortex by fast oscillations and response by slow oscillations was confirmed in humans (Buzsaki et al., 2013) but should be typical in all mammals. Down spin decoherence decreases the degrees of freedom; for example, through long-term potentiation (Bliss & Lomo, 1973). The increasing information content forms powerless, automatic actions.

As material fermions, the mental energy of temporal fermions and their field curvature develops in concert, thus they mutually determine each other (Figure 1.7.d). When the mind and the field are incompatible, emotional reaction is triggered. As the mental energy changes and adapts to the field, emotional reaction seizes. Repeated activation of the same neuronal connections requires less energy, resulting in less and less emotional involvement, forming automatic activation expressed by Hebb's law (Hebb, 1949), and hedonic adaptation (Schultz, 2007). Both examples clearly demonstrate the effect of the changing TF curvature on the mind. These processes give the cortical mind an immense advantage to adapt to environmental changes, to learn, and to form intellectual abilities. Therefore, *the brain can only restore its energy-neutral state on a different TF curvature by accumulating energy or information!* By changing mental energy, the mind (brain) remains congruent with its constantly changing environment.

Summary

Sensory information is transformed into a language of oscillations in the brain, which can be processed by the cortex. The complex electromagnetic flows in the intact brain are finely regulated by the constant and even minuscule changes in electric currents and potentials. The brain frequencies generate electric currents in the brain that are analogous to the motion of a spring. MiDT or MaDT are emotions, which dictate actions that recover the energy-neutral mental state by switching between harmonic standing waves (thus accumulating energy or information). Low oscillations (MaDT) flow information away from the cortex (energy flow to the cortex) and expand energy toward the environment through positive emotions and open, trusting behavior. High brain frequencies (MiDT) transport information to the cortex. The electric flow is extinguished by activating cortical neutrons on the cortical surface. The resulting electromagnetic imbalance reverses information flow by slower frequencies and recovers the energy-neutral state of the brain. As the activation of the necessary neuronal paths stabilizes, subsequent similar stimulus elicits only muted response as the energy-expensive emotional involvement is gradually replaced by an automatic action and the mind adapts to the TF.

SPINOR OPERATION OF THE MIND The section „The anatomy of interaction" in Chapter 1 discusses the spinor operation of material fermions. Low-entropy and spinor operation is exclusively connected to near-Euclidean environments, where the MiDV and MaDV maintain a relative balance. Societies also form low-entropy conditions characterized by competition and willingness for interaction and dictate spinor operation on emotional fermions. By orienting between the past and the future, the mind interprets stimulus as a binary code, either past (safety) or the future (the new). The mental states also form as either positive, or negative attitudes. But the stimulus and the mental state interacts to form a complex, nonlinear regulation, so the response of cerebral animals cannot be easily predicted. Depending on our expectation or attitude, the same stimulus can produce diametrically opposing results, the hallmark of spinor operation (**Figure 2.3**). The inverse sine function of Figure 1.4 expresses the changing curvature of the SF between the poles („The Heuristic foundation of hypothesis" in Chapter 1. However, since emotional fermions form orthogonally to material fermions, the names of the axes in Figure 1.4 should be reversed). Hence, the inverse sine function shows the changing curvature of the TF between the poles, and a positive TF is typical of societies.

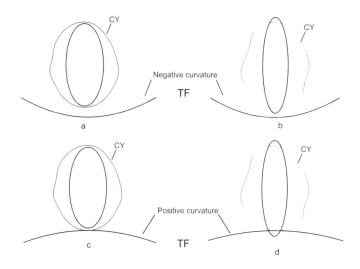

Figure 2.3. **Spinor operation of the mind** Decoherence depends on both the field curvature and the energy state of the particle, leading to spinor operation. Positive stimuli, low-entropy conditions, beauty, and love decrease the TF strength. Negative stimuli, high-entropy conditions (such as disorder), and irritation correspond to greater field strength. (a) The effect of positive stimuli on a positive-attitude mind leads to up spin decoherence, and negative field curvature. (b) The effect of positive stimuli on a negative-attitude mind leads to entanglement. (c) The effect of negative stimuli on a positive-attitude mind also forms entanglement. (d) The effect of negative stimuli on a negative-attitude mind leads to down spin decoherence, increasing field strength (forming positive field curvature). The opposing energetic changes occurring between the field and the mental particle in conditions (b) and (c) maintains the Euclidean field. The mental energy and the field curvature remain parallel in panels (a) and (d), which forms curving fields.

INTERACTION WITHIN CURVING FIELD As the reader will recall from „The anatomy of interaction" (Chapter 1), within a curving field a fermion's wave function returns to its original state after a 360° rotation (see Figures 1.2 and 1.3). Every interaction enhances the field curvature, so that the behavior becomes more and more deterministic (Figure 1.1 and 1.4). Since the poles limit the occurrence of Euclidean-event frequency, the mind has a hard time to remain neutral. Emotions constantly move us toward temporal polarities, happiness leads to happiness and aggravation generates anger. Entropic changes occur according to a U-shaped curve in material systems and in the mind (Figure 1.10). The expanding, up spin mind exchanges information for entropy, whereas in the contracting mind information and entropy increases parallel (**Figure 2.4**). This apparent contradiction occurs because mental states of pure quantum energy and of pure manifold energy are both high-entropy configurations (Figures 1.13 and 2.4).

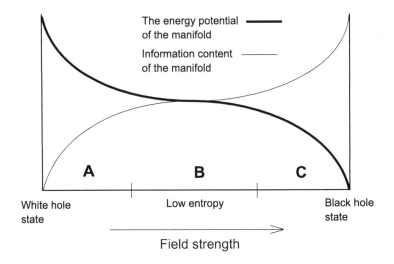

Figure 2.4. Relationship between mental energy and mental entropy High-entropy conditions are the poles, the black hole and white hole states. (A) The white hole state is typical of children. It is free of memory, its maximal energy potential shows up as trust, belief, and optimism. (B) The black hole state corresponds to the information-saturated (memory-saturated) manifold. (C) The middle portion of the curve is the low-entropy state. Starting from (B) and moving left on the graph (toward the white hole) results in the release of information for entropy in the expanding, up spin mind, whereas moving right (toward the black hole) results in information and entropy approaching each other in the contracting mind.

Within a positive curvature TF, the down spin configuration becomes the minimum-energy formation (Figure 1.8.c), and every interaction loses mental energy and freedom (see „The penalties of quantum energy"). Negative stimulus can take the form of flickering light, strongly delineated patterns, mechanically repeated noises, or criticism which enhances brain frequencies and fractures focus. The painful, constricted feeling of negative emotions not only feels bad but also seems to last longer, as time *perception lengthens*, leading to impatience, and stress (Fredrickson & Joiner, 2002; Yamada & Kawabe, 2011). Amazingly, this is true even for unconscious stimuli (Oei, 2012). The inhibition of associative representations narrows focus and reduces the degrees of freedom, so people cannot even see the possibilities open to them (Lupien et al., 2007). The lack of mental energy reduces trust and leads to uncertainty, turning ambiguous situations into down spin decoherences, or failures. The affected mind produces frequent interactions by criticism and aggravation (the Weyl tensor), which dumps quantum energy (MiDT) into the environment. The aggravation intermittently relaxes the mind, forming excessive kindness and accommodation (a Weyl-tensor effect). Although emotional experiences enhance memory, the detail-oriented attention of high brain frequencies corresponds to highly subjective memory, fractured by details and lacking overarching support. As a result, memories of negative experiences are contextually unstable and can shift their meaning. The distorted memory allows negative experiences to be repeated in essentially the same way over and over again, making it ineffective to instigate change. The mental life is shallow, repetitive, inflexible, involving schematic mental associations of partial, biased thinking.

In contrast, up spin decoherence is typical of negative TF. The MaDT decreases TF strength, so within negative-temporal-curvature layers the minimal energy configuration is up spin (Figure 1.8.d). Negative temporal curvature produces confidence, trust, and belief, which turns ambiguous situations into successful, positive outcomes (see „Mental expansion," below). For example, the willingness to work hard for rewards, even in cases of low probability of payout, is typical for a mind with high mental energy (Treadway et al., 2012). The energy exchange of one interaction is immense, but unlikely, because great mental energy forms trust, a noninteracting mental state. In mammals and birds, this trusting state eliminates the Pauli exclusion principle, allowing mental oneness through mating and caring for the young. In its higher form, romantic love in humans enables the formation of families. Correspondingly, a number of laboratories have shown that positive attitude—happiness—is the source of long-term wellbeing (Fredrickson & Joiner, 2002; Csikszentmihalyi & Hunter, 2003; Diener & Chan, 2011). Others found that being in the present moment or experiencing awe makes time seem longer (Rudd et al., 2012; Vohs & Schmeichel, 2003; Neupert & Allaire, 2012; Csikszentmihalyi & Hunter, 2003). As clocks slow due to gravity and acceleration, time *perception elongates* within both positive- and negative-curvature TFs! In temporal fermions, the positive-curving TF leads to more frequent interaction, but the energy change in each interaction progressively diminishes, which elongates the perception of time. In contrast, negative temporal curvature enhances the energy transfer of one decoherence, but makes it unlikely. In addition, decoherence reverses the direction of energy flow (negative time), increasing mental energy, which also elongates the perception of time. The elongated perception of time within a negative- orpositive-curving TF forms starkly opposite experiences. Within a positive TF curvature, excessive details overwhelm and exacerbate irritation, whereas within negative TF curvature, the wealth of time leads to mental calm and joy, forming emotional, mental stability.

Summary

A stimulus can give rise to opposite outcome in the temporal elementary particle, depending on the mental spin direction (attitude), and leads to spinor operation. Within a curving TF, the temporal fermion's wave function returns to its original state after a 360° rotation. Interaction changes the TF curvature and mental landscape. Increasing mental energy decreases the TF strength (leading to trust, confidence), whereas accumulating information increases the TF strength (forming uncertainty). A positive mindset turns an ambiguous situation into a positive outcome, whereas a negative attitude turns even positive stimuli into failures. The Weyl tensor increases brain frequencies, which narrows focus and leads to detail-oriented, contextually unstable memories, which are readily distorted. The detailed focus culminates in criticism, negativity, and aggravation (Weyl tensor), which betrays the eagerness for interaction. As a result, the positive temporal curvature forms a conservative, stiff world, which becomes increasingly disconnected from reality. Within positive temporal curvature, the minimal-energy formation becomes the down spin.

Up spin decoherence increases manifold energy. It leads to confidence and trust, which encourages up spin decoherence. It reduces the willingness for interaction, leading to mental stability. Noninteracting mental states form trust of love. Thus, sexual union (a noninteracting relationship) leads to birth, and the nurturing care of the young in emotion-forming animals. Both positive and negative temporal curvature enhances the perception of time, which reveals their connection to the changing field curvature.

EVENT-RELATED POTENTIAL (DECOHERENCE) The event-related potential is a family of wave components, occurring in response to stimuli. I propose that the event-related potential constitutes discrete, transient neural states (experience), thus a recognition or identification of the successive cinema of mental changes (Gorno-Tempini et al., 2004). For example mistakes form down spin state, which increases frequencies and therefore the focus, so corrections become possible. Short-term memorization requires a lot of energy. Recording the information liberates the mind, by forming up spin (Storm & Stone, 2014). For example hesitation is entanglement. Action ends entanglement and builds certainty, a liberating and empowering mental state. In traditional literature, an important part of interaction is the observer, which is always one of the two interacting particles. Hence, observer effect is the collapse of the wave function recorded by the event-related potential.

Over our lifetime the TF curvature increases, reducing the difference between harmonic frequencies, and corresponding event related potential. The smaller change in energy in one decoherence limits the brain's ability to absorb and comprehend external changes. In the elderly, the amplitude of one component of the event-related potential positively correlates with cortical thickness and executive function (Fjell et al., 2007), allowing us to stipulate a relationship between cortical thickness and mental energy in older people. This forms a fascinating chicken-and-egg puzzle: does the better mental functioning in the elderly protect the cortex or does a greater cortical thickness sustain mental energy?

Summary

The event-related potential is a family of wave components that are measured over the parietal and orbitofrontal lobes and is related to emotional processing, which might indicate the collapse of the wave function. When a stimulus unsettles the mind, it causes the wave function to collapse, which is the observer effect. A thin cortex might reduce the ability to absorb new information, thereby curtailing memory and reducing the event-related potential. Thus, the amplitude of event related potential in the elderly is found to positively correlate with cortical thickness and executive function.

Conserved current

Conserved current is a particle's refusal to interact, so energy is forwarded unchanged. Surprisingly, the mind can give rise to the same phenomenon. Conserved current is an automatic, easy, calm, and congruent reaction, which leads to the assured brushstroke of the master, the carefree but precise movement of the pianist's fingers over the keyboard. Conserved current represents the lowest energy—a stationary connection by automatic, unconscious action. Therefore, it does not involve emotion and does not generate new memory. This is why artists do a poor job explaining their own work. These instant, automatic decisions and actions are minimal-energy formations; therefore, they save energy (and time) and leave the mind energetic and powerful, permitting generosity and cooperation (Rand et al., 2012). In contrast, deliberations and delayed decisions activate conscious energies and lead to positive temporal curvature (Eimer, 1999; Gaal et al., 2012). The emotional mind's tendency to conserve current and decoherence separates over time. Our expertise and other behavioral comfort zones (i.e., confidence) ensure high entropy, within which we operate with ease, so we therefore tend to produce conserved current. Correspondingly, the competence of older adults was shown to parallel their confidence or self-reliance (Neupert & Allaire, 2012).

Summary

Conserved current is a minimal-energy formation of a relaxed, confident, or expert mind. It is a noninteracting state, which saves energy and time over conscious action.

Heisenberg uncertainty principle

Primitive animals with a limbic brain display a fairly predictable, linear regulation. The evolution of the cortex turns the mind into an emotional fermion, which obeys the Heisenberg uncertainty principle. Our emotional biases make us fallible, so we fail to follow classical logic, when it comes to attitudes and emotional judgments. The source of uncertainty can be found in the structure of the cortex. The manifold retains a memory; its response is heavily influenced by past experience. The response's nonlinear nature becomes especially prominent with enhanced stimuli, which produce polarized and even extreme responses.

The Heisenberg uncertainty principle prohibits the position and momentum of the particle from being known simultaneously. In the mind, the opposing poles of uncertainty are the *temporal position* and the *extent of emotion* (MiDT). Down spin decoherence uncovers only the extent of emotion, anger, or negative mood (how far one is willing to go). As with a cocked gun, which easily fires, decoherence only triggers the transported MiDT—an already-existing, accumulated, or pent-up anger or sadness, which originates in often unknowable past experience. In contrast, up spin decoherence uncovers the temporal position. As detailed earlier, positive emotions bubble up in the present moment and make happiness a perishable commodity, which cannot be stored up for another day. However, the extent of joy is a pointless question: we can only be fully happy.

Summary

The evolution of the cortex turns the mind into an emotional fermion. The memory potential of the cortex allows experience to guide behavior, leading to uncertainty. In addition, incentives produce polarized responses. According to the Heisenberg uncertainty principle, the position and momentum of a particle cannot be known simultaneously. For the mind, the two poles of uncertainty are the temporal position and the extent of emotion. Down spin uncovers the extent of aggravation or sadness, but its origin is concealed. In contrast, up spin reveals the time as the instant of interaction but keeps the extent of happiness unknowable.

Pauli exclusion principle

The Pauli exclusion principle states that fermions cannot occupy the same quantum state. For matter fermions, the principle is valid in space, but for temporal fermions (animals and people) it prohibits temporal closeness and generates **emotional distance**. It leads to territorial needs, or to avoiding eye contact in the elevator. Because the Pauli exclusion principle is responsible for the structure of matter, it also creates the structure of society or of ecosystems. It forms temporal entanglement of a common wave function that coexists between daughter particles over time in a quantum limbo, until wave-function collapse separates it into its energetically polarized particle pair.

The almond-shaped structure of the amygdala is equipped with extensive connections to a host of brain structures, which gives it a central and powerful role in emotional reactions and fear conditioning (Dolan, 2008). By regulating emotion, it controls behavior and memory, often outside of conscious awareness. Being activated by perceived proximity and emotionally charged images (even if we are not consciously aware of them); the amygdala regulates personal space and boundaries by moving us closer to others or farther from them. People with high mental energy or low **emotional temperature** tend to be satisfied and happy. Their mental calm makes them flexible and accepting toward others. As in colder matter, the Pauli exclusion principle is muted. The opposite is also true. In nervous, stressed individuals (with increasing emotional temperature), the Pauli exclusion principle and the critical tendency are strong. However, critical tendency only applies to emotionally close situations (i.e., temporal closeness). For example, eye contact shortens emotional distance. Therefore, to evade conflict, anxious people instinctively avoid eye contact. That eye contact enhances the potential for conflict between partners in conversation was confirmed by Chen and her colleagues (2012). This has no effect in the classical situations of everyday life; in big cities millions of people get along without major disturbances. However, in emotionally tight situations the opposing attitude becomes dominant. By profuse use of the word „no," even very young children attempt to separate themselves from the people closest to them. Of course, separation occurs in the emotional (conceptual) space of the mind, over a topological distance. Over time, loving partners and families actually tend to become more distant, and emotionally distant people (if spending time together) tend to approach each other emotionally. When faced with increasing emotional distance, we intuitively move closer in an attempt to maintain the emotional distance (Lenz's law applies for temporal fermions). Fans are attracted to celebrities, politicians, and saints. Their presumed distance permits an emotional closeness. This also makes it easy to be friendly to strangers and to tell secrets on the internet.

In oppressive class systems, the **societal entropy** is low. Historically, hierarchical societies maintained a low emotional temperature by the immense emotional distance created by the class system. As with matter, societies and ecosystems are also regulated by the second law of thermodynamics. However, in contrast to material systems, without outside interference (such as wars or natural disasters), the emotional distance between people decreases over time as order, democratization, culture, and civility increase.

Summary

The Pauli exclusion principle prohibits matter fermions from occupying the same quantum state and forces the temporal separation of emotional fermions. It forms the structure of matter and of society (or an ecosystem). The Pauli exclusion principle originates in the amygdala, a structure regulating personal space and emotional distance. The need for separation is more pronounced in nervous, stressed individuals, whereas it is muted in calm people. In hierarchical societies, low emotional temperature is maintained by the enormous social distance of the class system. In ecosystems and societies, the second law of thermodynamics increases entropy and permits democratization and the decrease of social distance over time.

The Elementary Particles of Emotion

Grouping temporal fermions

Leptons and quarks are the two classes of matter fermions. Leptons are further divided into neutrinos and electrons. As discussed in Chapter 1, the different physical and chemical characteristics of particle classes results from the energy differences of the torus. The emotional counterparts of matter fermions are temporal neutrinos, temporal electrons, and temporal quarks. Their differing energy levels betray an evolutionary progression of neural organization toward emotional complexity. The matter Calabi–Yau torus is presented in Figure 1.6.b. Through interaction with the SF, material fermions evolve and give rise to orthogonally oriented **temporal fermions (Figure 2.5)**, whose specifications are discussed in more detail in the section „Temporal elementary particles" of Chapter 3.

Three generations of particle families with vastly differing masses form each group of elementary particles. Their different masses connect them to environments separated by immense distances, whereas temporal elementary fermions are separated in time. Animals with greater **emotional mass** (temporal mass) always emerge in earlier evolutionary eras, and animals appearing later in evolution have smaller temporal masses. Today, only the lowest-mass temporal neutrinos (animals with a limbic brain such as sharks) and temporal electrons (mammals and birds) exist.

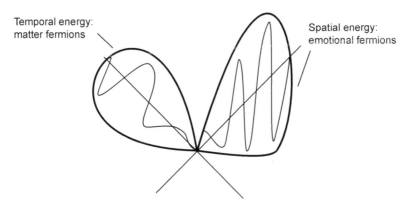

Temporal energy: matter fermions

Spatial energy: emotional fermions

Figure 2.5. Matter and temporal fermions within the universe The primordial oscillations gave rise to matter fermions (Chapter 1). Interaction of matter with the SF is a temporal evolution that forms temporal fermions.

The distinct character of fermions is determined by the torus energy. Temporal neutrinos, electrons, and quarks exhibit an evolutionary progression toward regulatory complexity (discussed in detail in Chapter 3). Each group of fermions can be divided into three families with increasing masses. Among the temporal particles, the three particle families are separated by evolutionary time. Organisms with greater temporal mass emerged first. It appears that today's environment is populated exclusively by the lowest-mass temporal neutrinos and temporal electrons.

Emotional forces

The evolutionary potential of the universe increases order and complexity. Evolution is a cosmologic process that insulates spatial energy within microdimensions by forming the temporal Calabi–Yau torus, the mind (Figure 2.5). The myriad specific mental phenomena can be intuited as being the emotional equivalents of gravity, electromagnetism, and the strong and weak nuclear forces.

TEMPORAL GRAVITY   Matter (and life) exists where expansion and contraction forces form slight gravity and the mind (and life) relies on temporal gravity. Just as matter cannot be shielded from the SF, the mind cannot be shielded from the energy of time. Every interaction adjusts mental energy. The socioeconomic layers of society correspond to the layered gravitational structure of the material world.

    The minimal-energy configuration of temporal fermions within temporal proximity is to have opposing spin (Figure 1.8.b), as dictated by the Pauli exclusion principle. Entanglement changes both particles' orientation to the field: the mental energy of entangled particles will change in opposite ways, thereby ensuring energy conservation. The temporal gyrocompass strives to recover its orientation against the TF. As explained in the section „Spinor operation of the mind," recovering the energy-neutral state is only possible by sacrificing (or gaining) mental energy, which actually changes the *mind*. In this way, the mind adapts to the curvature of the local field. The curvature differences of TF curvature reveal differences in trust and emotional sophistication (financial, social, cultural distinction) even in democratic societies (**Figure 2.6**), which explains why conflict and brutality accompanies poverty (great

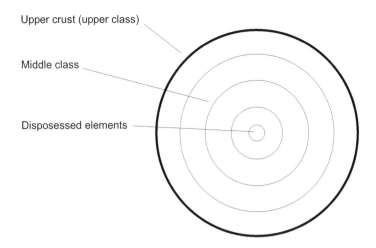

Figure 2.6. Structure of society Interaction separates MaDT and MiDT and generates the various temporal-curvature layers of society. The societal entropy increases. MaDT accumulates in the upper crust of society. The middle class occupies the central layers, where the equal balance of MaDT and MiDT lends great stability. MiDT accumulates within the layers with greatest curvature, causing the dispossessed to constantly struggle for survival.

TF curvature) and why civility normally coincides with members of the upper classes. However, individual temporal curvature is a question of attitude and it would be great oversimplification to associate temporal curvature with financial means!

During childhood, temporal weight forms in harmony with the home environment, and people strive to reconstruct this as adults. The socioeconomic status is stable when mental energy is congruent with the temporal curvature of the environment. People rise or sink in society (moving across the TF curvature layers), until their mental energy balance is compatible with the environment. Mental energy (optimism, trust, and hope) increases with education and work experience and dictates choices and behavior that lead toward success. However, mental energy generally decreases after middle age as the mind succumbs to egoistic limitations and accumulated negativity. This is why mature people are less trusting and increasingly focus on the past and preserving their status, which leads to actions and behavior that actually decrease their societal status and influence.

The powerful effect of temporal gravity (i.e., temporal weight) can cement abusive relationships. The temporal weight of the dominant partner increases over time in conjunction of the partner effort to keep the TF stable. This places a constant emotional pressure on the partner. This abusive relationship (in which one partner is bound within the constant hardship of increasing temporal limitations can become an emotional, energetic need for both parties. Emotional gravity (temporal gravity) can bind people (or animals) to the most sinister, horrific, or despicable personal circumstances and result in shocking personality changes. Many long-term marriages are kept together by temporal gravity, not love. However, in the majority of cases, the partners' gravitational grip is mutual. Each partner dominates in a different area of a love–hate relationship. They circle each other like two massive galactic objects, moving around stable temporal orbitals of well-defined limitations of a common center.

Summary

Because gravity has great control over material interactions, temporal gravity regulates the living world. Every interaction changes the mind. Positive time accumulates information on the mental manifold, whereas negative time enhances mental energy. Enhanced TF curvature decreases the degrees of freedom, overwhelmingly detailed information leads to impatience and stress, whereas the smaller curvature of the TF is liberating and enriches opportunities. The young mind adapts to the temporal gravity of the home environment and strives to recreate it in adulthood. Temporal curvature rarely changes in adulthood; however, mental energy often declines later in life. Emotional gravity can hold marriages together and lead to abusive relationships.

RELATIONSHIP BETWEEN SPRING MOTION AND TEMPORAL GRAVITY   The identical energetic dynamics of spring motion and gravity were discussed in the section entitled „Relationship between spring motion and gravity" (Chapter 1). Temporal gravity transpires by the same mechanism. Over time, interaction increases the social and economic differences, forming the temporal gravity layers. This process is discussed in more detail in the section „Societal evolution" in Chapter 3. The bell-curve distribution of temporal-gravity layers in economies and societies has been confirmed (Koonin, 2011). Therefore, temporal gravity is also described by Hooke's law. The constant k describes not only the stiffness of the spring but also the TF strength, which changes over time and defines a specific social organization (a historical time). For societies, gravitational relationships increase social differences over time so, in consequence, spring tension pushes away from the equilibrium point and increases field curvature differences.

Summary

Gravity forms the concentric, gravitational layers of the SF or TF. The particle quantum and manifold-energy distribution, being congruent with the corresponding field curvature, defines its position within the gravity field. People occupying the innermost TF layers of society form mental-energy-poor, „older" minds, compared with the regions having smaller field strength. Thus, if individual positions within the gravitational layers were represented as a function of time, a bell-shaped distribution would result, analogous to the gravitational distribution of matter. The relationship is described by Hooke's law.

MENTAL EXPANSION   The section titled „Acceleration" (Chapter 1) discussed the important differences between gravity and acceleration for material systems. As the reader will recall, stability requires compatibility between particle mass and spatial curvature. When the field strength is smaller than the mental energy, then excessive degrees of freedom leads to aloofness, superficiality, and emotional emptiness („moving with the tide"). Positive TF curvature causes information accumulation, leading to **mental contraction**, which is typical during aging. However, as excessive SF curvature prompts acceleration; immense TF curvature leads to **mental expansion** and increases mental energy. Mental expansion is the temporal equivalent of acceleration in material systems.

Physical discomfort and aggravation constitute the Weyl tensor that enhances brain frequencies. However, orientation toward a goal reduces temporal weight (the pain of temporal pressure) and transforms MiDT into MaDT, which is mental energy (Figure 1.14.b). In experiments with rats, the formation of new hippocampal maps was accelerated as a result of goal-oriented activity (Dupret et al., 2013). In human subjects, motivation to obtain rewards reduces conflict-related activation, thereby enhancing performance (Ivanov et al., 2012; Padmala & Pessoa, 2011); thus goal-oriented efforts are closely related to intellect (Wissner-Gross & Freer, 2013). These ideas are supported by neuroimaging data (Dixon, 2010). Immediate gratification reliably activates limbic reward regions, such as the ventral tegmental area. Limbic activation contracts the mind, whereas cortical activity expands it. Even self-forced patience, to remain non-reactant by accepting the irritation, suppresses limbic activation in favor of activation in the prefrontal cortex.

Csikszentmihalyi introduced „flow" in the seventies as an expression to describe the almost childish joy of the creative mental state.

The difference between insight and analytic solutions becomes clear when they are compared by using functional magnetic-resonance imaging (fMRI). Analytic problem solving entails increasing frequencies through focus and concentration, but brain frequencies actually decrease during creative insight (Kounios & Beenan, 2009). Therefore, inhibiting the natural analytic-problem-solving instinct (i.e., high brain frequencies) by artificial means, such as using low frequency brain stimulation, Lustenberger and her colleagues were able to significantly increase creativity in healthy subjects (2015). During creative solutions the degrees of freedom increase, leading to openness, trust, and confidence. Unnecessary details are eliminated and mental focus widens, allowing natural solutions to emerge. Creativity leads to positive emotions, but only a positive mindset can be creative (Isen, 2001; Fredrickson, 2003). Increasing mental energy forms negative field curvature, which deflects negativity and conflict, leading to creativity, success, and even longevity (Aknin et al., 2012; Steptoe & Wardle, 2005; Danner et al., 2001; Fredrickson & Joiner, 2002; Diener & Chan 2011; Stellar et al., 2015). So happiness and joy generate more of the same. Calm minds cannot easily be disturbed. Just as energy flows from warmer matter to colder matter, emotionally more stable (or „colder") people benefit from the negative energy, irritation (such as the Weyl tensor) of their environment and achieve mental expansion. Human history was shaped by this creative, ingenuous mental force.

Summary

Emotional interaction gives rise to temporal gravity and down spin decoherence leads to temporal confinement. However, acceleration permits one to escape the constriction of temporal gravity and achieve mental expansion. Focusing on a goal permits acceptance, which transforms the most difficult circumstances into inspiration and mental freedom (i.e., trust, belief, and the ability to see clearly). Quantum energy transforms into positive emotions (temporal excess). Correspondingly, analytic thinking involves higher brain frequencies, but the brain frequencies actually decrease during creative solutions.

EMOTIONAL ELECTROMAGNETIC FORCE   Electromagnetic force plays out in space (see Figure 1.16), but emotional electromagnetism modulates motivation over time and results in emotionally biased partial thinking. Maxwell's equations describe not only electromagnetism but also the changing nature of emotional electromagnetism. In the down spin state, the angular momentum of a quantum wave leaks out in the form of electric charge and leads to both gravity and electromagnetism! Emotional attraction is easily recognized (e.g., a dog wags its tail). Repulsion is revealed as the hair or feathers standing up on an animal's back. The extreme down spin state of emotional electromagnetism is incapacitating fear.

The charge varies according to a periodic harmonic motion, leading to emotional highs and lows. During its temporal movement, the orientation flips, turning attraction into aversion. The energy is constant during movement, although it changes from potential energy to kinetic energy and back. Being felt even in the stomach, these mental states can inspire or lead to physical consequences, such as disease. The emotional spring can only be controlled at the crests and troughs of the cycle: down spin states are energy starved (emotionally depressed) and up spin states are motivating. However, the highs and lows of the curve are the times when emotions are the strongest, so we have no incentive to change them. Mechanical energy moves the emotional spring between the extreme points of movement, and during this time emotions are hidden from conscious awareness. Attraction is a good example of emotional electromagnetism, where emotional distance (from the subject of attraction, such as a lover) changes over time according to a harmonic motion, moving between love and contempt. Emotional distance and emotional closeness are perfect emotional mirrors of each other and terminate in equal emotional distances from the emotionally neutral position. Deviation from the neutral position is unstable: increasing emotional distance intensifies attraction and decreasing emotional distance leads to doubt and uncertainty, seeking emotional distance in separation. Emotional electromagnetism obeys Lenz's law: it induces a mindset whose intentions and attitude oppose the original emotional force or charge. Belittling another will generate a feeling of inferiority, glee will engender jealousy, and desire will lead to pride. Admiring heroism in others will engender heroic behavior. However, the temporal expanse of emotional electromagnetism makes its effects difficult to recognize.

Summary

Emotional electromagnetic force continuously modulates motivation and results in emotionally biased, partial thinking. Emotional electromagnetism is described by Maxwell's equations.

The charge forms emotional highs and lows, following harmonic motion, which operates over time. Down spin states are energy starved (forming negative motivation) and up spin conditions are inspiring. The energy is constant, although it changes from potential energy to kinetic energy and back. Emotions are strongest at the crests and troughs of the cycle, so we have no incentive to change them. Between the extreme points of the curve, emotions are hidden from conscious awareness and change according to kinetic motion. Lenz's law applies for emotional behavior over time.

EVOLUTIONARY WEAK NUCLEAR FORCE (TEMPORAL WEAK NUCLEAR FORCE) As the reader will recall from Chapter 1, only suitable mass forms a stable connection with the SF. The weak nuclear force breaks down the unstable high-mass elements. In the living world, the **evolutionary weak nuclear force** spurs the evolution of organisms with smaller temporal mass. Every temporal elementary-particle family (emotional neutrino, electron, or quark) first formed with overwhelming quantum energy, which was manifested as an awkward and crude bodily organization (a rudimentary genetic makeup). The great temporal angle of the evolutionary weak nuclear force reduces temporal mass (increases mental and bodily potential) in a stepwise manner. Hence, evolution consists of long and stable ecosystems interrupted by short spurts of immense change. Gradual changes in temperature and habitat can push a species with high temporal mass into an ever-smaller living space, until survival is first challenged and then becomes impossible. Organisms with greater temporal mass (i.e., the species most dependent on the environment) succumb to evolutionary upheaval. However, organisms occupying smaller temporal curvature are spared. In a decimated, fragmented, segregated environment, boundary conditions represent the evolutionary potential of negative field curvature. By decreasing genetic entropy, the dynamic conditions stimulate rapid evolutionary change in surviving species (Szendro et al., 2013).

TEMPORAL MASS AND TEMPORAL WEIGHT Temporal mass is inversely proportional to the organizational sophistication of the genes, which determines the evolutionary stability of the organism (species) in a given environment (see Chapter 3). **Temporal weight** is the ego's emotional reliance on the environment. It is individual and situation dependent characteristic; Organisms forming great emotional weight are highly dependent on their environment for emotional stability.

Summary

The temporal weak nuclear force is the guiding force of evolution: it reduces the strength of the TF by reversing the temporal curvature. Every family of temporal fermions formed at first with a crude bodily organization and high temporal mass. Environmental changes challenge the survival of species with the highest temporal mass and lead to their extinction. Evolution spurs the formation of species with smaller temporal mass and greater bodily organization. Renewal of the decimated environment increases biological complexity.

TEMPORAL STRONG NUCLEAR FORCE   The strong nuclear force holds quarks together inside protons and neutrons. It is the strongest of all elementary interactions. The strong nuclear force is not only responsible for holding hadrons together but also flows over them and ensures the stability of the atomic nucleus.

In the animal world, the strong nuclear force has been the last and (very likely) strongest emotional force to form. Its short range indicates its connection to the Calabi-Yau space. The overflow of temporal strong nuclear force contibutes to social stability. Today, technological innovations decrease the temporal distance between people in an unprecedented way. The emerging trust leads to congruent relationships. I suggest that the strong nuclear force allows emotional closeness through altruism, trust, hope, and faith, which are uniquely human emotions. The societal glue of altruism exists in children as young as 30 months (Aknin et al., 2012).

Summary

In the animal world, the strong nuclear force has been the last emotional force to form, and it correlates with fluid intelligence and emotional stability. It is the foundation of the congruent society of the future.

Examples of typical mental states

Like material fermions, the temporal elementary particle is regulated by the environment. As a result, it displays analogue behavior, which plays out over time. Some consequences of emotions in human behavior are detailed below.

AGING Decoherence makes time (material changes) irreversible. Life events are also irreversible; this is reflected in our beliefs, which form the inevitable facts of our minds. Aging (temporal change) occurs fastest in young matter and slows in older matter (this is why inner clocks slow in proportion to gravity). In the mental dimension, the sensitivity to time is greatest in the young and leads to maximal mental expansion or contraction (Figure 1.13 white hole). Decoherence leads to great entropy (energy) changes in the infant mind. The intense energy flows (and corresponding emotional states) such as the whining of an infant can only be induced artificially in the adult brain (by stimulating the anterior cingulate). Boundless joy and excruciating pain render intense the emotional life of children. However, the great trust of children keeps the frequency of interaction small. (In practice, emotional states cannot be automatically connected to biological age. The biological state of the mind determines the potential of what is possible, but individual circumstances can differ.) The heightened temporal sensitivity of the young mind leads to hyperactivity—children cannot sit still. The small TF strength (very small temporal gravity) lends mental flexibility. Young people tend to adapt well and adopt every fashion trend. Because temporal gravity is so small in young people, emotional electromagnetism leads to overwhelming emotions, such as love. The   dominance of temporal gravity (against emotional electromagnetism) in the elderly provides great stability, but environmental changes become more stressful (Lupien et al., 2007).

The mental energetic change also defines memory capacity. The immense changes in mental volume in the young mind permits fast learning. With the increasing information content of the subconscious, changes in mental volume diminish during decoherence. As a result, the accumulation of mental energy due to a single decoherence is too small to trigger and stabilize new neuronal connections, so several exposures might be necessary to cement a memory (Figure 1.13, black holes state). The frequency of interaction is high, giving older people the sense that time is speeding up. However due to decreasing energy amplitudes in the anterior cingulate, the emotional difference between joy and pain declines. In the young, the mental sensitivity to pain is immense, although bodily discomfort is typically minor. In the old the reverse is true: the body truly suffers, but the sensitivity to pain is muted. This reversal permits great emotional stability and could explain the sense of wellbeing in old people (Urry & Cross, 2010).

Summary

The change in energy due to decoherence is greatest in young people and leads to immense emotional associations and significant mental expansion or contraction. In the elderly, the small energy change reduces memory capacity and the difference between joy and pain, which might explain their emotional stability.

THE TEMPORAL AND SPATIAL FIELD Gravity has a powerful influence not only over inanimate matter but also over biological systems. The swings of the SF and the TF regulate our mood in an intimate way. Although the spatial (gravitational) oscillations affect the body and the fluctuations in the energy of time affect the mind, both generate an emotional rollercoaster. Throughout history and in all cultures, the transitions from spatial contraction to expansion and from positive to negative time have been recognized as pleasurable. Negative time is the energizing surprise of the new, which such an elementary need for living systems. It is no accident that we simulate this transition in so many ways. Even crying in difficulty or after a tragedy is such a process. The temporal constriction gives way to emotional release, the feeling of spaciousness. Roller coaster rides are popular not because they generate gravity, but because of the feeling of lightness following it. Gravity is simulated by an upward vertical motion. As the ride reaches its highest point, gravity gives way to a feeling of weightlessness, so the feeling of contraction is replaced by the joyful feeling of expansion. Enjoying a transition from contraction to expansion is present in infancy. Children the world over enjoy swings and rocking in a cradle—the experience of gravity and weightlessness, contraction and expansion. Even adults enjoy rocking chairs. Children's stories from ancient times to the present depict the transition from positive to negative time, from emotional tension to release. The hero suffers and, the greater his suffering is, the more enjoyable his glorification afterward. Adventure, horror, cliffhangers, and suspense operate on the same principle. We suffer through every aversive predicament and emotional tension (positive time), and the payoff at the end gives us the emotional expanse of negative time. This is also the secret to the success of the twenty-four-hour news channels.

The brainstem is a personal entropic sensor that reacts to the temporal changes of the environment (which are registered as entropy) by adjusting the mood. Thus, the attitude (the mental spin direction) of the organism is regulated by the environment. However, the mind determines its own sensitivity to environmental entropy; temporal gravity (temporal weight) is highly subjective and situation dependent. The child curiously moves forward in a new situation with excitement (this represents the energy of the new) or pulls back in worry or fear (representing positive temporal curvature). Elegant and ingenious studies in psychology clearly demonstrate the effect of environmental entropy on mood and behavior. Low-entropy conditions (order, beauty) correspond to negative time and produce up spin, the feeling of satisfaction, happiness, well-being, relaxation, and the excitement of the new. Interest and the body position are open, trusting. The excited dog smelling around is in search of the new. Enclosed monkeys are willing to pull a lever to take a peek at the outside world: the new. Although the new is an elementary need, overwhelming and fast-paced information such as flickering light, strongly delineated patterns, or repeating mechanical noises provoke stress and increase brain frequencies. The temporal tightness of emotional gravity constrict the mind. The corresponding emotions of down spin states are anger, negligence, fear, paranoia, running, freezing, and adherence to the past (Ramos & Torgler, 2010). Even the language describes fear and guilt as difficult and heavy (Day and Bobocel, 2013), as it will be examined in more detail in „The penalties of quantum energy."

Summary

Gravity and temporal gravity has powerful influence on biological systems. The transition from gravity into weightlessness is a physical pleasure, which is stimulated by roller-coaster rides and swings. A shift between emotional tension and release (positive and negative time) is an enjoyable surprise for the mind, and it is utilized by children and adventure stories. Negative time is the energy of the new, which becomes an emotional need for emotional forming animals.

EMOTIONAL TEMPERATURE AND EMOTIONAL PRESSURE   The inverse relationship between pressure and temperature in gas was recognized in the nineteenth century and led to the universal gas law. Surprisingly, the universal gas law also applies to the emotional behavior of animals and people. In ideal gases temperature is proportional to internal energy and emotional conflicts are a direct measure of emotional temperature (MiDT). In the same way that particle collisions create pressure in gases, emotional confrontations generate temporal pressure, which translates into interpersonal and societal tension.

A gas is confined in a space, but the thermodynamic energy of the mind is defined by temporal coordinates. Emotional pressure (such as losing one's job, a privilege, or important territories of a nation) invariably increases the emotional temperature (the degree of aggravation) of the individual, group, or even nation. The measure of emotional temperature is the magnitude and degree of negativity, the extent of sadness, criticism, sarcasm, anger, or physical brutality. The negative energies are just mental tools that are used in an attempt to stretch the boundaries of the temporal confinement. Negativity is the thermodynamic energy of the mind! Criticism and anger provoke retaliation and reactions from the environment; which actually maintains the thermodynamic energy (the temporal pressure or temperature) over time.

Matter exists in space, with its temperature fluctuating over time, whereas emotional temperature is connected to circumstances and situations. Personal connections have their own chemistry; we behave differently with different people and relationships retain their flavor. Love remains fresh in the mind even after decades of separation! Just as energy flows from warmer matter toward colder matter, MiDT (aggravation) flows from people with higher emotional temperature toward people with lower emotional temperature (i.e., calmer people). This way, people tend to take on over time the emotional temperature of their environment. However, mental expansion (see section with the same title) exaggerates differences in mental energy. Just as various pressures and temperatures can produce fascinating phases and other states in material systems, emotional temperature and emotional pressure can be the source of interesting behavioral patterns in society.[4]

Summary

The mind operates within temporal boundaries. The mind uses MiDT as a tool to test how far it can extend within its environment. Therefore, the magnitude or degree of critical tendency or aggravation is a measure of emotional temperature or emotional pressure. Emotional temperature depends on the situation or relationship.

EMOTIONAL HYSTERESIS   Hysteresis is a memory-dependent behavior dictated by the history of the system. It is typical of elastic and ferromagnetic materials, but hysteresis has also been observed in many other fields, such as neurology, histology, genetics, and game theory.

---

[4] Additional examples of the psychological consequences of mental function are listed in the notes at the end of this chapter.

Here, we discuss the temporal hysteresis of the emotional fermion (i.e., the mind). In simple terms, the stimulation threshold of emotional beings depends on their emotional history (the emotional temperature, or the degree of irritation). Wikipedia describes emotional hysteresis on its page about catastrophe theory:

A famous suggestion is that the cusp catastrophe can be used to model the behaviour of a stressed dog, which may respond by becoming cowed or becoming angry (Zeeman, 1976). The suggestion is that at moderate stress (region A), the dog will exhibit a smooth transition of response from cowed to angry, depending on how it is provoked. But higher stress levels correspond to moving to the region B. Then, if the dog starts cowed, it will remain cowed as it is irritated more and more, until it reaches the 'fold' point, when it will suddenly, discontinuously snap through to angry mode. Once in 'angry' mode, it will remain angry, even if the direct irritation parameter is considerably reduced.

Such behavior currently has no scientific explanation. However, the emotional temperature changes of temporal elementary particles (temporal fermions) would easily produce the above-described phenomenon (**Figure 2.7**). Without cortical insulation, emotional neutrinos are incapable of emotional behavior, such as emotional hysteresis. The irritation of sharks and bees, for example, is rooted in simpler causes.

Social groups exist as emotional ferromagnets, stretched out in time. The emotional magnetic field becomes a common emotional orientation in most situations, signified by thinking and behavior. This is the reason social groups tend to shop in the same stores, take similar vacations, vote for the same candidate, and hold the same views on many issues. The field (which has a temporal expanse!) directs individual **emotional charge** in almost every problem. The common emotional orientation

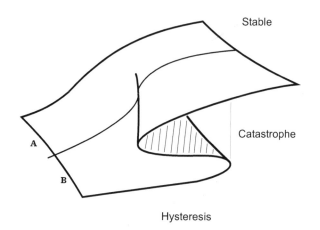

Stable

Catastrophe

A

B

Hysteresis

Figure 2.7. Cusp catastrophe model of emotional hysteresis
Regions with a single solution form stable behavior. In the folded region one unstable and two stable solutions are possible. Even small changes in a single parameter can cause a sudden change in the state of the system, which is called a „catastrophe."

changes gradually over time because of the experiences and emotional pushes of its individual members. A person's personal life depends so much on their social circle that individual response to challenges can be best predicted from the behavior of friends and peers! The shocking fact that social group is a more truthful predictor of a person's reaction to crisis than is personal intention is scientifically supported (Gilbert et al., 2009). Furthermore, behavioral modification spreads like an infectious disease within the social group—and this is true even between people who personally do not know each other at all! This finding came from social studies on happiness, quitting smoking, and discontent, but in all likelihood it may be found in many other behaviors and habits as well (Hill et al., 2010; Bliss et al., 2012).

Summary

The behavior of emotional beings depends on their emotional temperature, which is their degree of irritation. Emotional minds also display ferromagnetic behavior. For example, social groups form emotional ferromagnets, stretched out in time. The magnetic field dictates a common emotional orientation in most situations. Social studies support the idea that individual opinion and even behavior is greatly influenced by the social circle.

MENTAL INTERFERENCE   Interference occurs between coherent waves (equal frequency and identical or contrary phases). Constructive interference occurs when wave forms are in phase; waves that are out of phase subtract, which reduces overall wave size. Both material and temporal fermions display interference and resonance behavior. The context of judgments and decisions corresponds to interference in quantum theory. The presumed context of the first judgment or decision interferes with subsequent judgments or decisions, and produces a non-commutative relationship. The mind is an organic part of the   environment. As stimuli bombard us in a temporal rhythm, the inner   life of the mind also changes according to a wave form. Thus, mental interference occurs *instantaneously*, and without any conscious involvement. In the process, personal emotional tendencies are exaggerated or extinguished (added or subtracted from the temporal wave form of the stimuli). For example, degraded surroundings induces delinquent behavior (Keizer et al., 2008), and the smell of household cleaners promotes cleanliness (Liljenquist et al., 2010; Holland et al., 2005a,b).

Playing aggressive video games increases aggressive tendencies, but exposure to prosocial video games increases prosocial thoughts and behavior (Anderson et al., 2010; Greitemeyer & Osswald, 2011). Advertisements take advantage of this to coax our desire. Interference produces societal phenomena by way of temporal waves and bursts. Positive interference often leads to exaggerated interest, such as an investment bubble. However, over time, negative interference extinguishes enthusiasm and can even lead to avoidance.

Wikipedia gives a beautiful example about mental interference of temporal fermions. In their example, a person is twice given the chance to play a game in which they have an equal chance of winning $200 or losing $100. Whether they are told they won the first play or lost the first play, the majority of people choose to play the second round. Given these inclinations, according to the sure-thing principle of rational decision theory, they should also play the second round even if they do not know or do not think about the outcome of the first round (Savage, 1954). However, the majority of people do just the opposite in the latter case (Tversky & Shafir, 1992)—they do not play the second round if they do not know the result of the first. This and other, similar findings violate the law of total probability but can be explained as a quantum interference effect analogous to the explanation of the results of double-slit experiments in physics (see „Slit experiment, interference" in Chapter 1). Without feedback, the mind is in quantum limbo, waiting to form interference. However, information on the score would collapse the wave function forming the TF, which is knowledge; being liberated, the mind can move on. For example the process of grief is a formation of such an inherent knowing, leading to mental closure (see also the section entitled „Mind wandering").

Summary

Mental interference is a process in which energy is redistributed so that personal emotional tendencies are exaggerated or extinguished. Positive interference leads to exaggerated investment (such as financial bubbles), whereas negative interference extinguishes interest. Decoherence collapses the wave function, forming mental closure and eliminating interference.

QUANTUM ENTANGLEMENT   Entanglement is a phenomenon where by making an observation on one part of the system instantaneously affects the state in another part of the system, even if the respective systems are separated by space-like distances. The expanse of wave functions of matter particles over all of space permits quantum entanglement. The same phenomenon transpires in emotional fermions over conceptual distance, or over time. The confinement of positive temporal curvature leads to entanglement according to the Pauli exclusion principle. Entanglement is a common wave function divided between two sister particles separated by time, space or both. Hence entangled systems cannot be decomposed as separate subsystems. The semantics of concept combinations can be modeled by quantum theory (Busemeyer & Bruza, 2012; Pothos & Busemeyer, 2013). For example, in word association experiments entanglement activates associative target words simultaneously.

But entanglement can form between two interacting emotional minds. New information, even years later, can modify or update the energetic distribution between them, as dictated by their changing environment. One person can even form entanglement between two temporal events! With temporal fermions, the Bell nonlocality includes fictions time travel to the past that updates comprehension. The moment of recognition miraculously expands understanding in time, pushing it into the past and the future. In this way, childhood experiences can be viewed later by the mind of the adult. Analogous to the experiment of Scully and Druhl, a cemented mental reality can be completely overturned by new information (Scully & Druhl, 1982; Kim et al., 1999). When we discover a secret, our comprehension expands over time, and we are surprised that we could not see it earlier. The Bell nonlocality means that decoherence can be influenced from great spatial distances (for matter) or temporal expanses (i.e., for the mind). Ideas in the hidden corners of the mind can be manipulated by information years or even decades later.

Summary

The emotional mind can form entanglement over time. An entanglement is the separation of a single wave function into its opposite polar states: a complementary understanding that remains stable over time. New information can change the energy distribution between the entangled pair but cannot violate energy conservation. The separation of the common wave function ends entanglement. A single person can form entanglement between two time points, as new information changes her understanding in a retroactive manner.

MIND WANDERING (TASK-UNRELATED THOUGHT) During focused and detailed activity, the occipital lobe shows marked gamma frequencies (Kounios & Beenan, 2009), an energy expensive activity that necessarily limits its temporal expanse. A decrease in brain frequencies consequently fades conscious focus and impairs performance (Schooler et al., 2011; Christoff et al., 2009); activity shifts to the DMN (Mantini & Vanduffel, 2012), indicating inward thoughts of mental time travel. Such spontaneous disengagement from current activity is called mind wandering.

Using its free capacity, the mind is in a constant trial-and-error mode to find its limitations and possibilities (to discover the topography of the mental landscape). This temporal quantum computation forms a congruent mental picture from spatially and temporally disjoined and disconnected experiences. This automatic and nonconscious process formulates the direction of mental spin, which is the belief and attitude toward every question, and the basis of automatic mental activity. It leads to inherent „knowing." Recent studies (Smallwood et al., 2009; Fell, 2012; Mrazek et al., 2012) indicate that mental wandering is more frequent in people with a negative mood and that they also have more trouble reorienting to their task. In addition, mental wandering was found to result in unhappiness. Analytic thinking in tiring and can only lead to logical conclusions, limited by the initial boundaries of the problem. However, in the expanding mind quantum processes, such as quantum walk can lead to creative, radically new associations, ideas, and discoveries and produces immense happiness. It suffices to think of the purposeful and rewarding mind wanderings of creative geniuses, such as Mozart or Einstein, for example.

Summary

Mind wandering is a spontaneous disengagement from current activity, leading to curtailed performance, mistakes, and even unhappiness. However, in a creative mind, spontaneous mind wandering can give birth to new, creative ideas.

EMOTIONAL (TEMPORAL) FLUORESCENCE Incoming photons can boost the energy level of bound electrons and push them into higher, unstable orbits. During fluorescence, the absorbed energy is released (in the form of visible light) as the electron returns to a lower-energy state. The mental counterpart of this process is temporal fluorescence.

In psychology it is called passive aggression because it transports MiDT (criticism, aggravation) between people. Fluorescence can occur only in materials with bound electrons that can move into higher orbitals. Likewise, emotional fluorescence presupposes an emotionally bound state (such as a stable partnership), which allows the mind to form emotional distance but does not permit full separation. In the first step, a negative experience increases the quantum frequencies (MiDT), producing a feeling of hurt. In the constricted mind attention is turned to the past, in a detail-oriented focus, harvesting bitterness and aggravation. In the second step, the excess quantum energy radiates out in the form of criticism, sarcasm, judgment, innuendo, and irritable behavior. As the mental energy relaxes back to its lower-energy state, the mental focus moves back to the present and the relationship remains stable. Emotional fluorescence can operate between couples as a pendulum, moving MiDT back and forth in synchrony, identical to resonance fluorescence. Emotional fluorescence can also migrate from the presidential chair down the ranks, until someone kicks an innocent dog.

Summary

Fluorescence is not decoherence, just temporary energy enrichment. Just like its material counterpart, emotional fluorescence presupposes a (emotionally) bound state. Experiencing emotional pain, hurt increases brain frequencies and restricts focus, which is used to interrogate the past for problems, thus fueling aggravation. At the apportune time, MiDT is projected back out into the environment, as anger, criticism or sarcasm, formulated to perfectly suit the situation. In psychology, the phenomenon is called passive aggression.

LEARNED HELPLESSNESS   Just as matter relies on space, the mind relies on time for interaction (see section entitled „The spatial and temporal field"). Without the energy of negative time (new) mental progress is inhibited. This makes the new an elementary need! Sensory processing does not necessarily mean interaction, but interaction necessarily changes mental energy and forms a mental landscape. Just as walls or mountains constrict material movement, a partitioned mental landscape confers impassable *mental* limitations, whereas the negative-curving temporal landscape opens the mental horizon toward opportunities. The mental limitations demarcate the perceived possibilities of the mind and feel just as real, or more so, as fortifications made of rock and stone.

Approaching these perceived mental boundaries will trigger powerful emotional states, making violation impossible. Being an inherent part of the psyche, such a fenced-in mindset can be seen on the sad, resigned look of caged animals. Learned helplessness is the invariable, cruel end product of abusive relationships. The abuser forms immense TF curvature (temporal weight, see section „Mass," Chapter 1 and „Temporal mass and temporal weight"), its massive temporal choke gradually and irrevocably stifles the spirit of its victim. However, the abused also grows to rely on the emotional hold of the abuser. Thus, abusive relationships operate in a tight interdependence (see Depression). Martin Seligman named the phenomenon „learned helplessness" (Overmier & Seligman, 1967; Seligman & Meier, 1967).

Summary

The sensitivity of the brainstem structures forms a mental landscape, which determines behavior and thinking. A positive-curving field forms mental limitations, which demarcate the perceived possibilities of the mind. The resulting self-constructed limitations serve as a mental prison, out of which escape becomes impossible.

DEPRESSION (EMOTIONAL BLACK HOLE)   Depression is emotional isolation characterized by profound sadness, fatigue, and a feeling of worthlessness and guilt. It does not affect people when their sphere of influence is increasing but rather when their sense of confidence is challenged. Depression can be considered a mental death with the body fully functioning. The gravitational collapse of the TF forms the impenetrable event horizon of an emotional black hole. Excessive duty or expectation forms emotionally draining interdependence. The precondition of depression is massive temporal weight (gravitational curvature). This idea is supported by two works of Pulcu et al. (2014a and 2014b). The first work (2014a) found reliable enhanced amygdala activation in response to shame only in people with a history of depression. **Shame** is the eradication of security (the collapse of the temporal gravity field); therefore it indicates the enhanced emotional insecurity of large temporal weight. The second work (2014b) found enhanced altruistic behavior in guilt-ridden people. However, insecurity or guilt naturally leads to overcompensation by generosity, which is not true altruism.

We have seen that interaction is inhibited at the poles, the white and black holes (see „Entropy within the universe," Chapter 1). Mental energy gives confidence that leads to trust, and love; a situation in which interaction is inhibited (nothing can disturb the inner calm). This is the polar opposite of the depressed condition, which is an *incapacity* for interaction. The inability for openness in the depressed state is marked by enhanced self-referential processing of negative stimuli (Yoshimura et al., 2010). Time perception elongates, and mental progress is inhibited. Characterized by emotional rigidity, negativity, and increased sensitivity, depressed people strive for complete isolation. Except for occasional negativity (the emotional equivalent of Hawking radiation), emotional and social connection to others ceases. Due to gravitational dependence, the end of an abusive relationship can lead to depression. Note that depression does not occur in sustenance societies because survival requires constant interaction, which keeps the TF Euclidean.

Summary

Depression is characterized by excessive curvature of the TF, which encloses the mind and leads to complete isolation. The mind forms an emotional black hole, impenetrable from the outside.

THE PENALTIES OF QUANTUM ENERGY   Faulty emotional regulation leads to mental disease and madness. Aberrant thought patterns of psychiatric illnesses can often be traced to errors in emotional processing, such as social problems in the case of autism, persecution in paranoia, and unyielding conscientiousness in obsessive-compulsive disorder (Just et al., 2014). The possible role of emotions in mental problems gives urgency to better understanding of emotions, particularly the varied manifestations of overpowering quantum energy.

The hypothesis presented herein provides a coherent interpretation of numerous disparate studies by esteemed laboratories and scientists. As discussed earlier, enhanced conscious frequencies are important in many brain functions (sensory reception, analytic problem solving, motoric activity, and others). However, the persistent presence of MiDT signifies the chronic tendency for negative emotions, such as anger, irritation, or desperation. Enhanced quantum energy is the polar opposite of mental expansion. As mental expansion enhances mental power, leading to trust, the tendency for negative emotions siphons away mental power. The curving field reduces the difference between standing waves, curtailing the available mental energy during interaction. For this reason, over time, negative emotional states lead to insecurity, fear, and weakness.

Negative emotions might disappear from conscious awareness, but they manipulate the mental state from the background, corrupting mental abilities and blinding us to opportunities. As discussed in the section „Interaction within curving field" of this chapter, negative emotions lead to a *perceived* shortage of time. Therefore, MiDT handicaps focused work or analytic thinking and leads to negative arousal. This is supported by numerous studies on healthy brains by fMRI and positron emission tomography (PET). Anxious individuals are attracted to and spend more time with negative stimuli (such as looking at angry faces), which only heightens their aggravation (Bethell et al., 2012). A study by Van Dillen et al. (2012) found that participants display more severe moral judgment after experiencing negative emotion (disgust), even if the judgment in question is unrelated to the original emotion. Guilty people are insecure. For example, people with a guilty conscience have been found to overestimate their own weight and to consider chores to be more difficult than those with clear consciences (Day and Bobocel, 2013). The preponderance of negative emotions was shown to parallel high-frequency limbic activity and decreased prefrontal functioning (Seo et al., 2008). Indeed, efforts to suppress negative emotion increase activity in the prefrontal cortex while reducing activity in the amygdala and nucleus accumbens (limbic region) and giving hope that persistent conscious attention can lead to mental change. (These results are consistent with studies done on creativity see „Mental expansion.") Beyond the energy need of greater brain frequencies, negative emotions affect the entire body and result in long-term negative health consequences (Diener, 2011). Like the derailed train that moves inevitably toward an accident, emotional cruelty and negativity accumulate penalties on the field curvature (increasing TF curvature), connecting high brain frequencies and aggression (Ohman & Mineka, 2001). Over the long term, this state unfailingly gravitates toward failure, sickness, and moral justice. This long-term cause-and-effect relationship is recognized as karma.

Summary

Negative emotions hinder personal and professional success and over the long term can even lead to health problems.

FREE WILL   Matter particles are fully governed by the environment, but the brain's control over its own bodily real estate creates the belief in free will. However, the common belief that our life is governed by conscious thinking and intellect has been increasingly challenged. As early as 1983, Benjamin Libet questioned the existence of free will by showing that thinking and conscious actions are preceded by unconscious brain activity (see inset).

It is hard to fully appreciate the power of unconscious brain activity directing our lives. For instance, patients with damage to the visual cortex (V1) often show blindsight: they can correctly guess, when questioned about stimuli that they do not see (Weiskrantz, 1990, 1998). Even more remarkably, amnesic patients with hippocampal damage can discern between new and old stimuli without any conscious involvement (Laeng et al., 2007). Also, processing emotion-laden stimuli has been shown to be largely automatic, irrespective of the focus of attention and, in some cases, independently of visual awareness (Ohman & Mineka, 2001; Pessoa, 2005). For example, emotional faces evoke responses in the amygdala in spite of attention being directed elsewhere. Subliminal stimuli can influence the prefrontal cortex for a long time, even without eliciting a conscious response (Van Gaal et al., 2012; Custers & Aarts, 2010). Conscious decisions are preceded by measurable neural activity (Soon et al., 2008, 2013; Custers & Aarts, 2010; Aarts & Van Den Bos, 2011; Murakami et al., 2014; Bengson et. al., 2014) and neural indecision is accompanied with behavioral delay or hesitation (Kaufman et al., 2015). Voluntary decisions are often predictable based on ongoing brain activity. For example, the decrease of alpha amplitude or total suppression of alpha activity positively correlates with visual detection (Wyart & Sergent, 2009, Mathewson et al., 2014) and the momentary measure of alpha activity predicts where conscious attention will be shifted (Bergson et al., 2014). But brain activity and behavior can also be actively modified. For example stimulation of the ventral tegmental area, implicated in reward and depression, can change motivation in monkeys (Arsenault et. al., 2014).

The sluggish conscious decisions have vastly longer time requirements from the fast, automatic decisions. Conscious processes take a second or longer, but fluid mental operations are overwhelmingly automatic and occur in fractions of a second. Conscious focus also becomes quickly tiring, but the automatic mind operates over the long term and remains stable in the face of environmental changes. We have to consider that our automatic mind, highly influenced and regulated by the environment and operating behind (sometimes against) conscious awareness, determines the course of our lives. The mind forms a unified experience by connecting sensory perception with mental states based on event related potentials (Mancini et al., 2011; Guterstam et al., 2015). Whether action occurs due to priming by conscious or unconscious (subconscious) stimuli, the mind presumes its ultimate causative role, which is the belief in free will. Yet it is hard to appreciate the environment's ability to direct our lives through emotions. As discussed in the section entitled „Connection to Theory of Relativity, Quantum Mechanics, and String Theory" of Chapter 1, the spatial field unequivocally directs the behavior of material elementary particles, and the TF has inevitable power over our actions (Brady & Anderson, 2014). The fact that simple animals, such as rats, for example, can be directed by implanted electrodes seems to be a natural consequence of neural biology. Although to an outside observer they are clearly enslaved by their dependence, addicts and other substance abusers claim to be in full control of their lives. We have seen that conscious decisions lead to MiDT, which is the narrow and distorted focus of positive temporal curvature. Down spin states (selfishness, stress) distort mental vision and twist our memories. Accumulated details build obstacles, leading to wild mental swings, which is analogue to the shear stress of the Weyl tensor. The constantly changing attention eliminates freedom. As a result, people with negative attitude are enslaved by their circumstances and behave as puppets on a string. Their conscious minds are employed as public relation agents, to constantly explain away previous behavior. Free will is only possible for a mind with high mental energy, but this is a satisfied, trusting, calm, or happy state when there is no emotional incentive for change. The moment we decide to change, however, quantum energy increases, attention narrows, and free will is lost. The silver lining for mental change is goal-directed effort and acceptance of circumstances. We might play the leading role in our lives; but the play is written and directed by the universe. The perfect cinematic spectacle gives the illusion that we are in control.

Summary

Numerous studies convincingly demonstrate that conscious decisions are preceded by measurable unconscious activity. It has also been demonstrated that electrically stimulating the brain modifies conscious preferences and reliably produces certain behavior. This is exemplified by the rat that can be governed via implanted electrodes. The mental universe is overseen by its electromagnetic balances and resulting emotions. Automatic decisions, which occur in a fraction of a second, have great power to direct our lives. Quantum energy and MiDT enslave the mind to its circumstances due to the narrow, constricted focus they impose. Operational freedom is permitted for the calm or joyful mind, which has no incentive to change.

THE REGULATORY IMPORTANCE OF EMOTIONS   The regulatory importance of emotions in mammals and birds is overwhelming.

– Pain sensory nerve cells are profusely distributed throughout the body. The sensation of pain determines our physical limitations, but emotional pain sets our opportunities and mental limitations by framing our emotional comfort zone. Emotions regulate every aspect of life. Even the way we sit or stand is emotionally determined. There is never an emotion-free minute in our lives.

– Our emotions determine whether we want to live or die, but they do not let us to take our lives; suicide must be planned carefully.

– When motivation is strong; it can even give us superhuman abilities. Without motivation, we do nothing. The greatest emotions, such as love, faith, or pride inspire the greatest achievement.

– The behavior of mammals and people (and even nations) can be impeccably manipulated by emotions. This is intuitively understood by every great teacher, politician, or artist.

Conclusions

The brain can only restore its energy-neutral state over time by accumulating energy (or information). This turns the mind into a temporal gyrocompass.

I propose a new hypothesis for consciousness and mental operation. The mind forms a structure that is symmetric and oriented orthogonally with respect to matter. It is thus an elementary fermion, which interacts with the TF via elementary forces, the emotional equivalents of gravity, electromagnetism, and the strong and weak nuclear forces. Matter interacts with space to produce time; whereas the mind transforms time into mental volume, and forms mental energy. Thus, time and space can transform into each other. This hypothesis underlines and supports the interconnected and interdependent nature of existence. Both energy fields of the universe, space and time, are enclosed within microdimensions, limiting their transformations into quanta. We must consider that, like material systems, consciousness operates according to the principle of least action via elementary forces.

Electromagnetic activity in the brain is not arbitrary or accidental; it is directed by charge-conservation laws. The minimal-energy state of pairs of spinor fermions is, as dictated by the Pauli exclusion principle, for them to have the opposite spin direction, which is called entanglement. Shifting brain frequencies form temporal energy vacuum: emotions. In turn, emotions force actions that, through decoherence, change mental energy (and thus the field curvature). However, the mind is also a temporal gyrocompass, which always restores the mental energy balance (the ratio of quantum and manifold energies). In this way, the mind recovers its energy-neutral state and remains true to the local field. However, in the process, low brain frequencies accumulate energy, whereas high brain frequencies accumulate information. *Therefore, in its constant interaction with the environment, the brain constantly changes and adapts!*

The quantum operation of the mind and the importance of emotions are quietly gaining ground in a variety of disciplines; for example, in neurological, psychological, and common-sense observations. Traditionally, emotion is considered secondary to higher mental functions. Here a radical opposite idea is postulated: mental operation is based on emotion. Emotions are the exclusive motivating force of higher animals. Over time emotional interactions form a mental landscape, which regulates the sensitivity of the brainstem, thus its reaction to the environment. High mental energy leads to emotional stability. Behind every great piece of art, music, literature, or science we find positive emotions, for only the high mental energy mind is capable of creating, or „flow." Even art that describes terror and suffering carries a positive, optimistic emotional message: the belief in the human potential. And at the basis of abstract mathematics we find harmony, symmetry, and beauty. The temporal excess of positive emotions increases mental energy. Faith, love, courage, and awe bubble up with the enthusiasm of the instant. Without connection to the TF, the force cannot be transferred to tomorrow. Thus, MaDT is the force of the moment, but its long-term effect is profound. In the expanding mind, unnecessary details are eliminated, expanding the sense of time and fueling enthusiasm, generosity, and the energy for happiness and joy. It is important to note that excitement, glee, and selfish joy are not positive emotions because they increase quantum energy. At the same time, catharsis and reaching closure in sadness are uplifting and lead to mental expansion.

The MiDT of negative emotions, however, remains part of the mental landscape for an extended time. Detailed focus wastes time, distorts reality, and turns experience into a house of mirrors. Thus, the mind becomes partial and acts contrary to its own best interest, leading to regret and remorse. The back-and-forth emotional swings of the Weyl tensor do cancel but waste energy in the process so, by most measures, life suffers a gradual decline (Fredrickson & Joiner, 2002). Like the distractive turbulences of a fast-flowing river, conflicts destroy mental progress and inevitably produce failure. This hypothesis explains the mental costs of negative choices, the liberating nature of acceptance, and the power of goal-oriented activity (such as learning) in mental growth. The symmetric energy structure of matter fermions and temporal fermions makes it possible (with appropriate corrections) to transport conclusions and understanding between theoretical physics and neuropsychology, for the benefit of both. This short book is intended to form a first step toward building a comprehensive theory that describes the operation of the mind. I humbly invite the scientific community to consider and test the hypothesis. Any evaluation, contribution, or criticism from interested scientists would be appreciated.

Notes

Explanation of some sociological and psychological findings is given below.

THE NEGATIVE EFFECT OF ENVIRONMENTAL DISORDER ON BEHAVIOR In response to the high entropy of environmental disorder (such as uncontrolled trash piles) the effected mind forms positive TF curvature that increases brain frequencies and leads to stress and the corresponding asocial thoughts and behaviors, such as social phobia or stereotyping people. It also encourages delinquent, careless behavior or negligence (Ramos & Torgler, 2010). Reversing the effect of enhanced brain frequencies seems to be possible by brain stimulation (Sellaro et al., 2015). This clearly proves that the shift in electromagnetic balance caused the observed behavioral changes in the first place. As shown in the section entitled „The penalties of quantum energy" a down spin state (MiDT) is weighed down by its focus on unnecessary details, and lack of patience, which produces the observed behavior.

THE PERFORMANCE OF SPORTS TEAMS IS PROPORTIONAL TO THE FREQUENCY WITH WHICH THE MEMBERS TOUCH EACH OTHER The report by Kraus and colleagues (2010) show a positive correlation between the frequency with which teammates touch each other and team performance. Trust and emotional closeness between team members permit physical touch and spontaneous embrace, which is relaxing. Low brain frequencies allow the players to rely on their creative subconscious, which puts them in the flow. This is the secret of a fluid game.

MEN WHO PLAY CHESS AGAINST ATTRACTIVE WOMEN BECOME RISKIER PLAYERS Male chess players choose significantly riskier strategies when playing against attractive women (Dreber et al., 2013). Sexual arousal is a relaxed mental state, ruining conscious focus and concentration. Thus, the male chess players resorted to subconscious activity, relying on creative but risky playing tactics. Apparently, chess requires training, concentration and analytic focus. Pleasure, especially sexual arousal, has been used successfully in spying since biblical times (remember Samson and Delilah). In the relaxed mind, high brain frequencies allow trust, which brings truth to the fore, and to the tongue—especially truth that was consciously held back. The high frequency required for conscious focus (short term memory) limits it for short-term (Howard et al., 2003). With the relaxing of conscious focus, secrets are divulged.

THE MORE PEOPLE DOUBT THEIR OWN BELIEFS, THE MORE THEY PUSH FOR THEM   When people are presented with evidence that undermines their core conviction, they tend to become more forceful and blatant about those convictions (Gal et al., 2010). People with high temporal weight vigilantly protect their stability (safety of the past and the status quo). As discussed in the section „Emotional temperature and emotional pressure" of this chapter, contradiction produces quantum energy that fuels emotional pressure and temperature. So conflict produces retaliation, which maintains the pressure. Proselyting, an automatic generator of contradiction becomes a handy tool for the mind laden with quantum energy. However, every unsuccessful attempt to convince others further erodes trust, producing a vicious cycle. People who feverously proselytize are desperately trying to maintain their high emotional temperature. For this reason their best subjects are not converts, but people who resist and retort their arguments. In fact, accepting the view point of such proselytes only prompts them into more paradoxical arguments to provoke continued confrontation.

THREE

THE ROLE OF ENTROPY
IN EVOLUTION

Abstract

Evolution is an inherent and sweepingly powerful organizing force in the universe. The Calabi–Yau torus, which is an energy-neutral structure, reoccurs on three different levels and turns the cosmos into a fractal. Interactions produce discrete changes in energy, and the inverse modifications of space and time lead to singularities, which are the poles of the universe. Black holes form in places of spatial contraction, whereas spatial expansion generates white holes. White holes decrease entropy and nurture complexity through evolution. Building on top of previous progressions, physical, chemical, and biological evolutions form subsequent phases. Physical and chemical evolution creates complex material structures and prebiotic chemicals. Organic stews nurtured by the entropic, thermodynamic pressures of their physical environment produced life. Biological evolution consists of very well defined periods that are separated by vastly different environmental conditions, culminating in the emergence of the mind and society. The apex of evolution is the appearance of the still-evolving, highly intelligent mind, which cannot be a uniquely human quality—other planets and other galaxies necessarily nurture life. The universe's drive toward complexity is a greatly humbling realization, because the most complex matter known to man is the human brain. Chapter 3 is divided into five sections: The first section discusses the evolutionary phases. The second section introduces the structure of the universe, whereas the third section introduces the temporal elementary-particle families, the laws governing temporal fermions, and the mechanism of social evolution. Section four is about the self-regulation of the universe, section five concludes.

"The increase of order inside an organism is more than paid for by an increase in disorder outside this organism. By this mechanism, the second law is obeyed, and life maintains a highly ordered state, which it sustains by causing a net increase in disorder in the Universe."

– Erwin Schrödinger

Introduction

After 13.8 billion years of purported entropy increase we are witnessing a universe putting on violent and fierce spectacles with incessant energy. The maxim of Rudolph Clausius that „the entropy of the universe tends to a maximum" rings hollow. However, as discussed in Chapter 1, the second law of thermodynamics is only true within gravitational environments. The accelerating expansion of the universe is a constant energy generator, which increases the entropy of the universe as a whole, but decreases the entropy of its constituents. In gravity-free space and within gravitational regions, opposite entropic changes take place. Evolution is an ever-increasing complexity that works through the self-organization of matter into stars, planets, and living creatures such as us.

Charles Darwin laid down the basic idea of evolution with great insight, but in light of the enormous progress since then in genetics and molecular biology, the basic premise that random processes, stochastic mutations aided by selection pressure, and survival of the fittest could have given rise to sensory, emotional, and intellectual complexity is now suspect (Merlin, 2010, Hedges et al., 2015). The emergence of DNA repair mechanisms has reduced the evolutionary importance of arbitrary mutations.

In addition, arbitrary mutations more likely lead to a loss of abilities than to the development of innovations. For example, domesticated animals were vigorously selected for specific qualities. Although their sizes and colors show amazing variety, a dog is still a dog and remains only a dog. The Darwinian idea of evolution can produce amazing variety in sizes, colors and shapes but, to create the complexity of fins, hearts, and the human mind, it becomes hard pressed for answers. The Darwinian theory of evolution is in a need of a serious update (Saphiro, 2011).

From elementary particles to emotional minds, ecosystems, and societies, we often find behavior that hints at a discrete energy structure (Wolfram 2002). When discrete structures interact, energy imbalances can be stabilized and enhanced, leading to structural differences, i.e., complexity. Modification of the field curvature generates congruent energetic changes within the Calabi–Yau torus. The system's behavior is not arbitrary; interaction produces a memory, which in turn becomes the source of further action, as shown by the Bayesian game theory (Harsanyi, 1967; 1968). The increasing entropy produces conditions favorable for the emergence of life (Wissner-Gross & Freer, 2013). Every unsuccessful attempt at life helped to channel evolution and made the environment more favorable to life. Biological evolution is not arbitrary; it inevitably advances toward the emergence of the mind.

The Structure of the Universe

We have seen that the two Calabi–Yau-space building blocks of the universe (matter and emotional fermions) differ by several orders of magnitude, and form vastly different energy levels. The universe's accelerating expansion is a continuous creator of space. As we saw in Chapter 1, cosmic voids have strict energy structures, so expansion generates energy. However, within the near-Euclidean gravitational regions of the universe, energy conservation is strictly observed. During interaction, MiDT in one emotional mind (or species) is balanced by MaDT in the other and the concentration of MiDT and MaDT remains in equilibrium in the universe, leading to charge neutrality (although local imbalances can form). Landauer's principle states the convertibility of energy and information, which also means the equality of energy and information. This can be recognized as the mechanism of „static" time. Proven by Moreva and colleagues, it means that the universe forms an energy-neutral unit: the physical laws are limited to and characteristic of the universe, which cannot be divided and from which nothing can escape. It is reasonable to suggest that the SF and TF represent extremely low-frequency standing waves within a Calabi–Yau torus, which is called the universe, which necessarily should be primary. The all-encompassing, self-regulating and intelligent universe gave rise to material fermions, and wherever the necessary minimal conditions exist, it engenders the emergence of the intelligent mind. Thus, the universe is stretched between the white holes and the black holes, and the expansion of the white holes is kept in check by the enormous field strength of the black holes (**Figure 3.1**).

Black holes

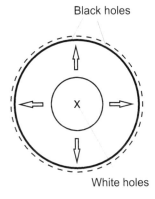

White holes

Figure 3.1. Structure of cosmos with white holes and black holes
The microdimensions of the cosmos form a closed minimal surface. The expansion generated by the white holes (indicated by white arrows) decreases the overall curvature of the universe, as shown by the dotted line. The galactic environments gradually absorb the expanding volume by building manifold area (MiDV). The outer boundary of space is formed by the contracting black holes; their great field strength stabilizes the universe and prevents runaway expansion.

Therefore, black holes form the edge of space. This is the ineviteble conclusion of Almheiri and colleagues (2015), who examined black hole entaglement and found their horizons impenetrable firewalls. In short of introducing some novel physics, their only viable alternative is a black hole horizon, which is the forbidding boundary of the universe.

The three layers of the universe's elementary building blocks, the material fermions, the mind, and the universe itself, are highly interconnected in spite of their enormously different sizes. However, matter is completely dependent on space for its operation; the mind can govern its bodily real estate, whereas the universe is self-contained and self-regulating. Thus, the three elementary particles represent increasing degrees of freedom and manifest increasing complexity. The identical structure of material fermions and the mind indicates their fundamental connection: matter originates at a temporal point; at zero time, whereas the mind originates at a spatial point; at zero volume. Material fermions form spatial complexity, whereas temporal fermions give rise to mysterious and inexplicable mental complexity. Matter fermions use up space to produce temporal evolution, culminating in the emergence of the mind. Mental fermions consume time to form mental expansion, the source of spatial (mental) volume. Thus, the orthogonal orientation of elementary particles (matter and mind) form a predator-prey relationship and embrace as yin and yang, determining each other's future and past. The structure of the Calabi–Yau torus, reformulating from the largest to the smallest scales, leads to a fractal and cellular structure formulated by submanifolds in material, biological and societal structures.

Summary

The accelerating expansion of the universe is a constant energy generator; however, the universe observes energy and charge neutrality. The energy structure of the Calabi–Yau torus reappears on three levels as the building blocks matter, the mind, and the universe itself. Its energy neutrality makes the universe an elementary fermion.

A new concept of evolution

James Maxwell's 1867 thought experiment involves a demon or a device that would select warmer particles and channel them unidirectionally in a divided container. The warmer channel would increase the temperature difference between the chambers and violate the second law of thermodynamics. Maxwell's demon, however, expresses the power of self-organization, which is the capacity and most essential feature of living organisms to counter and even make use of the environment's increasing entropy.

Evolution often moves forward in seemingly chaotic processes that nevertheless have an unescapable, directional flow. Evolution can be divided into physical, chemical, biological, and societal phases, where the names of the phases refer to their defining parameters. The sequential and increasingly complex steps are well separable and are characterized by specific temperature, pressure, entropic, and oxidative qualities. Physical evolution began at the high temperature of the Big Bang. Interactions produced the galaxies and the elements of the periodic table. Chemical evolution proceeds on dust and ice surfaces of low-temperature gravity-free space, where fast moving atoms freed after annihilation of stars readily assemble into complex molecules with biological potential. Life's nurseries are rocky and temperate planets with mild gravity. The narrow temperature and pressure range of Euclidean environments is ideal for biological evolution.

Darwin's idea was radical not only in his time; evolution continues to be controversial, particularly in relation to the second law of thermodynamics. Evolution's apparent drive for order is hard to reconcile with the second law. However, organization can be viewed as a tool that abets the entropy production of the inanimate environment. This was Schrödinger's insight. The possible deep connection between entropy maximization and organization, or even intelligence, has been hinted at in fields as diverse as computer science and cosmology. Maximum entropy production seems to describe nonequilibrium processes in physics and biology (Matyusev 2010; Wissner-Gross & Freer, 2013). The entropy production of gravitational environments is recognized by the second law of thermodynamics. I propose to call the entropy-maximizing drive of the gravitational environment **entropic force**, which forms an **entropic pressure** vis à vis membranous organisms. Entropic pressure is proportional to the entropic changes of the environment and inversely proportional to the relative concentration of membranous organisms. Entropic pressure becomes the source of evolutionary change which, assisted by elementary forces, slowly accumulates useful energy (order) inside the membrane and gives rise to self-organization—the emergence of life and evolution. The entropic pressure of the environment is maximized in dynamic systems where the relative concentration of membranous organisms is small. Typically, this can be found in transition states, boundary conditions, or in species with small or fragmented population size (Nevo & Beiles, 2011). On the surface of Earth, energy infusion (such as the energy of the sun) increases chemical turnover. Gravity (i.e., the second law of thermodynamics) and the continuous, steady energy input became the driving forces of biological evolution. The physical laws govern not only the nonliving world but also the biological systems. Recognizing evolution as a fundamental science governed by physical laws can revolutionize our ability to manipulate and manage biological systems.

Summary

Evolution can be divided into physical, chemical, biological and societal phases. Gravitational environments' entropy maximizing drive is an entropic force. This force forms an entropic pressure, proportional to the entropic changes of the environment and inversely proportional to the relative concentration of membranous organisms. On Earth, gravity and the energy input of the sun drive evolution toward increasing complexity.

Evolutionary phases

The first phase of evolution: physical evolution

The unmeasurable chaos called the Big Bang involved violent pulsations that calmed into regularity and formed the Calabi–Yau space, which became the structural foundation of space and the universe. The nearly-scale-free variation of temperature of the CMBR leads to today's uneven spatial structure in which spatial contraction with extreme gravity is balanced by regions of expanding cosmic voids. Star formation and annihilation continuously enriches interstellar space with elements, particularly heavier elements, such as oxygen, nitrogen, carbon, sulfur and others having biological potential. Therefore, their concentration in interstellar space is an important marker of the galaxy. The greatest concentration of heavy elements, about 2% to 3% is thought to be a requirement for life. This concentration is normally found in the youngest stars, such as our sun. The location of the star within a galaxy is also presumed to be important, because radiation levels and gravitational perturbations would make the central region dangerous for life. A quiet interplanetary neighborhood of a lower-mass star at the outer region of a galaxy is thought to be preferable for biological evolution. To form a moderate climate, the planet needs to rotate asynchronously with the parent sun, so heat is distributed evenly throughout the atmosphere (Leconte et al., 2015). There is also speculation that the planet must have a large satellite, which provides regular, mild tides and generates plate tectonics. It also would ensure a stable tilt of the planet, giving rise to seasons, which are thought to be an evolutionary advantage. The search for extrasolar planets, even with our relatively crude current technology has yielded several rocky planets in temperate orbits in our own galactic neighborhood. About one in five sun-like stars are estimated to have Earth-like planets in the habitable zone. Earth's experimentation with life is certainly not unique. The number of habitable planets (and biologically evolving) throughout the universe is probably staggering (Bovaird, 2013).

Summary

The Big Bang formed the structural foundations of today's universe, where regions of gravitational contraction are balanced by regions with spatial expansion. The stars, galaxies, and elements of the periodic table were formed by billions of years of physical evolution. In the youngest stars, such as our sun, the concentration of higher-mass elements has reached the level necessary for life. In the vastness of the cosmos, an immense number of planets should exist that satisfy the special requirements necessary for biological evolution.

The second phase of evolution: chemical evolution

The ubiquitous presence of complex organic compounds in the near-vacuum environment of space, not only within our galaxy but throughout the universe, is a puzzle (Kwok & Zang, 2011; Öberg et al., 2015). The expanding universe pulls large galaxy formations away from each other, but leaves regions tightly connected by gravity in a contracted state. Such volumetric changes can form a highly reactant environment. Chemical potential is the greatest where fast flowing heated plasma (star matter) is subjected to sudden expansion that occurs when crossing from regions with great gravity into expanding space and collides with cold free-radicals. As exploding stars spew high-mass elements out into gravity-free and decreasing-entropy interstellar clouds, ultraviolet radiation, other radiation, and magnetic fields produce free radicals on icy dust surfaces. Radiation and sudden temperature and pressure changes can catalyze orderly and complex organic reactions (Pizarello, 2010; Kwok, 2009; Kwok & Zhang, 2011; Ehrenfreund et al., 2011). For example carbon, the fourth most prevalent element in space, tends to form long carbon chains and aromatic rings of great structural variety in the presence of high concentrations of hydrogen and water. Kaiser et al. (2013) found proteinogenic dipeptide formation by electron irradiation in interstellar ice analogs. Although most organic matter ends up as fuel for the formation of solar systems, traces do remain interspersed in cosmic debris and meteorites. Their isotropic composition corroborates that their organic complexity originates in decreasing-entropy, gravity-free space. Although close to 200 different organic compounds have been catalogued in meteorites, thousands more remain unidentified. These compounds, such as fullerenes, purines, amides, large aliphatic aldehydes, acids, aromatic compounds, and others are probably representative of the great natural organic variety of stellar environments (Hudgins, 2002). Due to their great stability, carbon polymers of aromatic and straight structures, such as kerogen, when packaged inside comets and meteorites can easily survive intact during travel to the far corners of the galaxy (Pizzarello, 2006; Pizzarello, 2010). This organic richness supports the hypothesis that stellar organic compounds can kindle biological evolution wherever the necessary minimal conditions are satisfied. For example, the Late Heavy Bombardment around 4.1–3.8 billion years ago impacted the inner planets of our solar system in particular. NASA's GRAIL mission found a lot of fracturing from massive impacts during the first billion years of the moon's life, which also should be true for the inner planets, and especially for Earth. Meteorite impacts must have been fairly regular events that destroyed much of the early crust, but they may have provided raw materials for a dynamic and highly responsive environment that favored experimentations toward life.

Molecular chirality is an unmistakable and essential quality of living systems, so the chiral L-asymmetry of amino acids of some meteorites has puzzled scientists. However, several studies hint at possible mechanisms for their interstellar synthesis. L- and D-enantiomers can transform into each other when subjected to a strong magnetic field, which means that chiral transformation could occur in the aftermath of star annihilation (Bordacs et al., 2012). Marcellus and colleagues (2011) were able to reproduce in the laboratory the asymmetric formation of amino acids upon irradiation by ultraviolet light within a presumed interstellar environment. The unique L-asymmetry of extrastellar organic compounds might have provided the template for the L-asymmetry of archaea glycerol, leading to the chirality of biological systems (Pizarello & Shock, 2010; Pizarello, 2006).

Summary

Thousands of organic compounds have been observed in space. They can form in interstellar clouds, where high-atomic-mass elements spewed out by exploding stars could interact with free radicals. Temperature and pressure changes and radiation exposure could catalyze orderly and complex organic reactions. Carbon compounds packaged inside comets and meteorites can easily survive intact during travel to the far corners of the galaxy. There are also indications that strong magnetic fields or ultraviolet light can induce chiral transformations. These extraterrestrial chiral compounds may have provided the template for biological chirality on Earth.

The third phase of evolution: biological evolution

EMERGENCE OF LIFE (CHEMICAL AND BIOLOGICAL EVOLUTION) The Earth continuously gains energy from the sun (the net energy gain is the difference between the incoming and reflected radiation). The reductive atmosphere of early Earth likely consisted of some combination of carbon dioxide, hydrogen, methane, water vapor, and ammonia. Without a protective ozone layer, magnetic radiation and ultraviolet light would have wreaked havoc on all molecules, forming reactive ions, free radicals, and plasma. The large organic compounds of meteorites could be hydrolyzed into smaller, complex compounds with biological potential. The temperature was high compared to today, so fast, energetic chemical reactions would have brought forth an imaginative array of compounds, many of which do not exist in the natural environment today. Processing organic matter by hot, acidic electrolytes generates small, reactant compounds, which form the raw materials for the relentless entropic organization of organic compounds by the dynamic environment.

Today, organic content is quickly decomposed or eaten. The biological food chain and oxidation keeps the open waters clean. On primeval Earth, however, without atmospheric oxygen or primitive life forms, the seas could accumulate and recirculate organic compounds freely. Radiation made the reductive, watery milieu (near or at the boiling point) highly reactive; in particular, the solubility of moderate-length hydrocarbon chains increased as a result of nitrogen, phosphor, sulfur, and oxygen becoming incorporated into the molecules. Small hydrolytes could remain suspended indefinitely in the reductive aqueous solutions due to hydrogen bonds and other ionic interactions. As Earth cooled, this rich organic content formed colloids, emulsions, and suspensions in pools of hot water. In the hot electrolytes, amphiphilic molecules (having both water-loving and fat-loving ends) formed prodigiously. The increasing entropy in the slowly cooling waters encouraged micelle formation, because joining the hydrophobic ends of the molecules frees water. The incidence of low-entropy micelles allowed the system's overall entropy to increase, because the order inside the micelles was balanced by the increasing entropy of the environment. Cacophonic organic variety, together with mechanic and electromagnetic influences, such as electrostatic discharge by lightning, could also have produced liposomes and other membranous systems.

In small pools of water the continual hydrolysis of large, insoluble organic molecules into smaller, soluble compounds formed an entropic pressure. Once the concentration of micelles became saturated; the entropic pressure could drive organic molecules inside micelles or other membranous systems. Hydrophobic membranes prevent the transfer of inorganic (hydrophilic) ions, but facilitate diffusion of organic compounds and ions. The slowly increasing volume gradually expands the membrane structure, forming a membranous vacuole. Dynamic changes in pH, temperature, electrolyte balance, and chemical composition would have initiated a wet-cell battery of ion-concentration differences between the two sides of the membrane. The changes in pH due to hydrothermal vents, for example, could result in regular voltage differences across such membranes, which would drive amphiphilic compounds inside (or out of) the membrane. The hydrophobic interior within the membrane would have encouraged simple organic ions to combine into larger molecules by releasing water molecules. Because the resultant larger molecules could not cross the membrane, complex chemical organization would be favored inside the membrane, while entropic conditions would favor the breakdown of large molecules in the environment. Organic ionic movement across the membrane could form sinks (accumulations of positive charges) or sources (accumulation of negative charges) of electric charge, resulting in a passive ion flow. Since elongated membranous systems form greater charge separation, such elongated shapes could accumulate greater structural complexity. Under the hot, acidic conditions, the entropic drive of the environment, over time, would store and accumulate energy in reversible reactions inside the membranous vesicles and enhance complexity in a slow and measured fashion. Molecules with allosteric properties could form reaction cycles with increasing stability.

In the slowly cooling water the size of membranous systems would increase, which would also expand their inconstant and highly evolving membranes. Mechanical forces, geothermal activity, outbursts of hydrothermal vents, weather changes, and ocean currents would have stirred the water and agitated the membranous systems (organic sacks), rocking and mixing their internal contents. Oversized sacks inevitably would break in two or dissolve into the environment, preventing them from growing excessively. Although simple organic sacks would not be capable of consumption, they could maintain a stable, low-entropy environment in their interiors (homeostasis) and could multiply by primitive division—breaking up. Reversible reactions would store energy in organic compounds, inadvertently turning the thermodynamic and chemical potential of the environment into engines of energy accumulation. Over time, in primitive vesicles the chemical processes stabilized, and size correction (division) became a regular, internally driven event. Therefore, the ability to maintain homeostasis and a low-entropy state, supported by division and consumption, is the primary requirement of life. These heterotrophic organisms can be considered protocells.

But the colorful experimentation of this era cannot be traced back, and not only because fossils could not form. Earth's early chemical experiments predated the emergence of hereditary material, where the organizational history of the organism accumulates. Analogous to the formation of the Calabi–Yau space as the beginning of time in the universe, the formation of genetic material is the beginning of biological evolution. Because hereditary material and protein structure evolved together, examining the genetic code allows us to draw conclusions on the structural character traits of early life. Imbalances between genetic structure and protein structure leads to correction, enforcing their parallel, orthogonal evolution and making genes the memory of evolution. Genetic material is thus the manifold (holographic history) of biological systems.

The reductive atmosphere and high temperature of early Earth dictated the special molecular composition of the membranous structures. Remnants of such primitive organization can be found in some archaea, where two polar heads are attached to two fused phospholipids. Thermophilic bacteria and archaea have membranes that contain stiff ether bonds rather than the usual ester bonds. In fact, archaeal membranes can be in the liquid-crystalline phase at temperatures ranging from 0 to 100 °C.

In addition, archaeal membrane composition is very stable and has extremely low permeability to solutes (Koga, 2012). Looking at the archaea (considered to be the most primitive life form), it is clear that their internal structure is less ordered (higher entropy) than that of any other known organism. For example, their membranes consist of isoprenes, which are highly variable in their connections and show an almost liberal physical structure. In addition, their membranes contain L-glycerol in contrast to the D-glycerol of bacteria and eukaryotes. The arbitrary change from L to D chirality sharply divided evolutionary pathways. Since L chiral cells were excluded from the archaeal gene pool, the small number of cells containing D-glycerol had a fresh start at evolution by capturing the entropic pressure of the environment. These D-glycerol organisms evolved into all the phyla known today. The genetic code is nearly universal in the living world, which indicates the remarkable economics of evolution and the interconnectedness of all species resulting from evolution from a common ancestor. A case in point: only small genetic differences account for the enormous morphological variety of mammals and a third of the human genome originated from bacteria.

Summary

Meteorites and comets provide a regular supply of water and organic material to planets. The oxygen-free atmosphere and reactive milieu allowed organic material to remain suspended indefinitely in the ancient seas. The increasing entropy in the slowly cooling waters encouraged micelle formation and the changing pH and entropic pressure of the environment drove amphiphilic compounds into their interior. Over millions of years membranous sacks would have accumulated complex compounds; these can be considered as protocells. Mechanical forces would regularly break these membranous sacks, preventing them from growing too large.

However, evidence of the crude experimentation of this era is hard to uncover because it predated the genetic record. Archaea, the most primitive life form, is endowed with the most primitive organization of the living world. Their membranes contain the more reduced ether bonds instead of the usual ester bonds and highly variable isoprenes and L-glycerol. The arbitrary change from L to D chirality excluded L chiral cells from the archaeal gene pool. By capturing the entropic force of the environment, these newly formed cells containing D-glycerol would evolve into all the phyla known today. Being nearly universal in the entire living world, the genetic code betrays the remarkable economics of evolution and the interconnectedness of all species.

Evolution of life

The overwhelming majority of organisms that live on Earth are considered simple by evolutionary standards and show such an excessive diversity—not fully understood even by scientists—that their classification is difficult. The two living kingdoms on Earth that most people are familiar with are animals and plants. However, the regulation of homeostasis remains simple in plants. Behind the seemingly limitless morphological variety—the colorful petals and intoxicating fragrances—lies a very simple organizational background. However, the anatomical structure of the digestive system or the complexity of the eye or of the kidney has no parallel in other kingdoms. The following discussion will focus on the evolution of animals.

The reductive, radioactive atmosphere and high levels of radiation of early Earth resulted in practically limitless genetic experimentation that sowed the seeds of life and produced the first living organisms. The basis of life's genetic diversity was born in that reductive milieu. It is believed that the archaeal lineage deviated very early in evolution. Archaea has several important ancestral characteristics that set them apart from all other life forms. With their unusual membrane structure consisting of L-glyceride and ether linkages, archaea represent a living „fossil" record of life's beginnings. Disadvantageous changes quickly disappeared due to mortality. Archaea and bacteria represent a flexible genomic pool, where clearly separated genetic differences form a compartmentalized gene structure (Koonin & Wolf, 2008). However, horizontal gene transfer allows random mutations to move about in populations.

The archaea and other ancient organisms became increasingly efficient in eliminating organic waste from open waters—the decomposition process. The emergence of life changed the chemical dynamic of ancient Earth. Organic compounds, which occurred freely in water, were actively absorbed by these organisms, which ignited a biological chain that neutralized and recycled all waste. With the emergence of cyanobacteria, organisms became able to sequester the sun's energy. Photosynthesis uses the energy in sunlight to break and form chemical bonds, producing energy-rich carbohydrates and freeing oxygen, a strong oxidizing agent, in the process. The slowly accumulating free oxygen in the atmosphere gradually oxidized metals and organic materials in the environment.

Oxygenation of Earth's atmosphere seems to have happened in two stages. After the first stage, the oxygen level reached only about 2% to 3%. Even at that level, the oxygenation of Earth's atmosphere revolutionized the environment. The anaerobic ecosystem was relegated to small pockets of oxygen-free niches of the habitat. The oxygenation of the atmosphere allowed some filtering of the ultraviolet radiation, which slowed the rate of genetic mutation. The cellular chromosomal repair mechanism had not yet evolved, or was primitive at best, so endless genetic experimentation could still occur freely, perfecting the cell structure. For example, mitochondria are thought to have evolved from a symbiotic relationship with bacteria.

Methane is a potent greenhouse gas. Its oxidation into carbon dioxide and water by free oxygen is thought to have caused of the long Huron glaciation 2.4 billion years ago. During the 300-million-year-long icy winter, life persevered. During the slow, cold survival years, metabolic pathways probably further stabilized. In the second oxygenation phase, the oxygen level reached or exceeded today's level of 21%. As the oxygen levels gradually advanced, the newly formed ozone layer shielded ultraviolet radiation, reducing the mutation rate and stabilizing genetic material. Higher-order protein structures, constrained by protein folding, stabilized protein sequences and led to globally conserved evolutionary patterns. The expansion and complexity of genetic material paved the way for multicellularity. Remarkably, most of the known 40–50 biological phyla, which is almost all of the extant organisms, originated in this oxygenated and slowly warming environment characterized by low radioactivity.

The two phases of oxygenation radically transformed the ecosystem. As discussed in Chapter 2 in the subsection of the same title, the life-supporting capacity of the environment can be renewed by the evolutionary weak nuclear force. For example in first phase of oxygenation the mass extinction of anaerobic organisms would have enriched the seas with organic material, supplying rich fodder for emerging aerobic life forms. At the same time the life-supporting capacity of Earth—due to the presence of oxygen—for complex organisms have increased. The second oxygenation could support the greater metabolic needs of the circulatory system and the sensory and digestive organs, spurring faster, larger organisms with greater bodily organization and mobility. Evolution appears to be a saga of increasing complexity and intellectual organization, from protein folding, to the mind. Homologous evolution is rampant, and proteins show remarkable versatility for new and unrelated functions (Morris, 2010). Maximal and opportunistic use of environmental resources leads to numerous examples of parallel (or convergent) evolution. For example, some twenty different linages autonomously developed multicellularity, locomotion was perfected in water, on land, and in the air and the wing evolved independently at least three times in multiple lineages. The evolutionary vacuum left after the extinction of the dinosaurs was filled in a blink of an eye by evolutionary standards by the evolution of mammals. The startling similarity of mammals that emerged independently on two or three separate continents shows that the interconnected complexity of the ecosystem is not accidental; it is abetted by the evolutionary weak nuclear force.

Summary

The great genetic diversity of the living world originated in the early genetic experimentation of the ancient Earth. Oxygenation and the developing ozone layer slowed the mutation rate, which allowed cell structure to stabilize. Animals have attained an exceptional degree of complexity through evolution. Remarkably, major groups of extant organisms belonging to the known 40–50 biological phyla originated in the aftermath of the 300-million-year-long Huron glaciation. Oxygenation permitted the evolution of faster, larger organisms with complex organs, such as the circulatory system and sensory and digestive organs. In an inalienable drive toward complexity, evolution uses all available recourses and takes advantage of all environmental opportunities to advance complexity.

Characteristics of an Evolutionary Era

As shown earlier, physical, chemical, and biological evolutions are described by very different parameters, and subsequent evolutionary periods are built upon previous ones to produce outcomes with increasing complexity. Biological evolution is a succession of distinct evolutionary periods forming eras and eons that are characterized by specific environmental conditions as well as the genetic and morphological characteristics of the organisms that populate them. Abrupt transitions, characterized by a major loss of species richness, disrupt ordered and stable ecosystems and result in a discontinuous fossil record. The subsequent evolutionary periods are found to build upon the genetic and environmental potential of the preceding era. The ecosystem gradually degrades its home environment, which makes survival difficult; for example, predator-prey cycles, and symbiotic relationships can fall apart. Species most dependent on environmental conditions are the first to become extinct and are followed by many others. Only small remnants of the ecosystem survive in the decimated environment. As the environment regenerates, animals and plants aided by the accumulated genetic potential of the previous era quickly spread and flourish. This energetic inevitability of the decimated ecosystem to move to a higher level is the evolutionary weak nuclear force. Indeed, the boundaries of geological ages have been marked by mass extinctions and have been followed by rapid diversification and the accelerated genesis of novel organisms (Saphiro, 2011). Some major extinctions include the early Cambrian event [540 million years ago (mya)], the Cambrian–Ordovician event (490 mya), the Late Ordovician extinction (445 mya), the Late Devonian extinction (370 mya), the Late Permian extinction (250 mya), the Late Triassic extinction (200 mya), and the Late Cretaceous extinction (65 mya).

Evolution is highly dependent on genetic diversity, the variation in the gene pool, and mental entropy, which is determined by the curvature of the TF as shown in the section „Interaction within the curving field" of Chapter 2. Population (societal) entropy is a statistical quality, formed by the mental entropy of members of society or ecosystem. As the reader will recall from „Entropy within the universe" of Chapter 1, high entropy occurs in either pure manifold energy or pure quantum energy situations (see Figure 1.10). In the same way, high **mental entropy** can be the sign of either the mental openness of trust (manifold energy), or of an agitated, insecure search for survival (quantum energy) see Figure 2.4. Individual and species entropy changes together and regulates mating behavior, which in turn determines **genetic diversity**. Genetic diversity is an evolutionary asset because it allows survival of the species in changing, adverse conditions, which is typical of evolutionary upheavals (see areas A and C in Figure 3.3). Mental entropy changes parallel to genetic diversity, but artificially regulated populations, such as domesticated animals, are exceptions. Because of artificial breeding, purebred dogs and farm animals show little morphological variety (low genetic diversity). Their genetic diversity is low, but their mental entropy is high due to their safe and relaxed environment. This is the reason that these populations exhibit low environmental stability. In other words, natural mating would sharply increase their genetic diversity. This is why wild dogs are mutts, as opposed to wild coyotes or wolfs. Genetic diversity and population entropy deserves a detailed analysis.

Summary

Biological evolution is a succession of distinct evolutionary periods characterized by specific environmental conditions as well as the genetic and morphological potential of the organisms. Evolution occurs by abrupt transitions; the destroyed ecosystem and a major loss of species are followed by quick diversifications and fast renewal. Evolutionary periods are built on the evolutionary potential of the preceding era. The environment supports the ecosystem but is gradually degraded by it. Organisms that are most dependent on the environment cannot weather the changes and die out. Thanks to the evolutionary weak nuclear force, the decimated ecosystem gives way to a new evolutionary cycle with a higher evolutionary potential and greater complexity.

Genetic diversity is an evolutionary asset because it ensures the survival of a species during evolutionary upheaval. Although pets and domestic animals show little genetic diversity, they form high mental entropy due to their protected environments.

STAGES OF EVOLUTIONARY PERIODS   Many notable differences exist between the material world and the living world. For example matter marches irreversibly toward black holes as space is gradually transformed into microdimensions. Ecosystems disintegrate by the same process, but the evolutionary weak nuclear force allows living systems to renew and to evolve toward order and complexity in a gradual process.

The cycle of death and birth is a constant source of change and renewal of the living world. The conclusions of Chapters 1 and 2 indicate that the structure of the Calabi–Yau torus determines submanifolds with energy-neutral structure on many levels. For example, the ecosystem forms an energy-neutral submanifold, ensuring the balance of nutrients, wastes, air, energy, and other resources over time. A perfect energy balance exists only in the Euclidean temporal curvature in the middle section of the cycle, whereas the beginning and the end stages are perfect energetic mirrors of each other. At the start of the period, nutrients are plentiful in the fresh ecosystem (forming entropic pressure) whereas, at the end of the period, disorder and waste accrue (increasing temporal gravity) and collapse the evolutionary period. The weak nuclear force allows the cycle to be repeated but with greater complexity. Accordingly, evolutionary periods can be divided into three stages, the first stage with profuse energy, the second with energy equilibrium and third (broken energy-nutrient cycle) stage.

An evolutionary period begins with few surviving species of the previous evolutionary cycle. Only small numbers of survivors populate distant pockets of a fractured environment, which is reflected in high genetic and morphological diversity. The TF curvature is negative, and the dynamically changing environment exerts significant entropic pressure on the small and fragmented populations. During this period the genetic entropy is actually decreasing! Therefore genetic changes are not arbitrary, but converge toward organization and functional utility. Quickly emerging genetic innovations allow genes and proteins to acquire new functions and find new uses, which can take advantage of the fast-changing environment (**Figure 3.2** and **3.3**, area A). New organisms and new species appear from almost nowhere in seemingly arbitrary evolutionary jumps. Bizarre morphologies and unexpected features of early evolutionary periods were noted by Morris (2010), among others.

The second stage of evolution (Figures 3.2 and 3.3, region B) is characterized by low entropy; the TF is Euclidean. Interactions between the species (competition, predator-prey relationships) are governed by the Pauli exclusion principle. Interaction always produces winners and losers—there is no neutral outcome! The differences in living space reflect the species' place within the temporal gravitational layers (Figure 1.11 and 2.6). Species occupying layers with smaller TF curvature have vastly better life prospects than those occupying layers with greater TF curvature. The number of layers defines the diversity of the ecosystem.

In the third stage of evolution (Figures 3.2 and 3.3, region C), the recycling of nutrients is unbalanced, the competition for resources strips the land of vegetation, and prey becomes scarce.

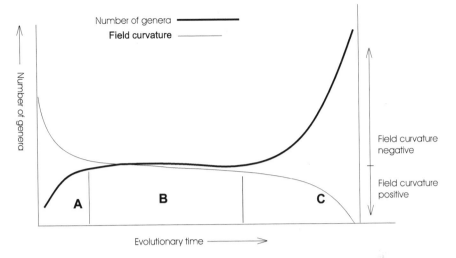

Figure 3.2. Example of biodiversity during an evolutionary period The curve can be divided into three regions A, B, and C: the first region A has negative field curvature. The Euclidean region characterizes the second stage B, and the third, final region C forms positive field curvature. Biodiversity increases in the first and third stages of the evolutionary period. However, in the first stage, decreasing entropy allows evolutionary innovations to emerge by the reorganization of genetic material, whereas in the third stage genetic changes spread in the ecosystem and increase entropy. The first stage is characterized by evolutionary jumps, while genetic refinements lead to morphological differences in the third stage.

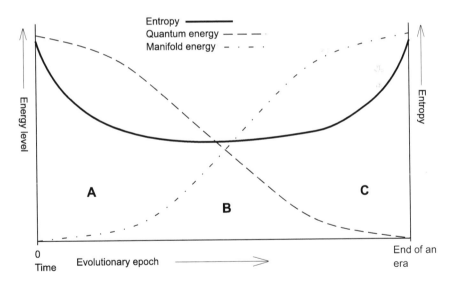

Figure 3.3. Changes of ecosystem within a single evolutionary era The ecosystem strives toward energy neutrality throughout the evolutionary cycle. (A) The evolutionary era is jump started by the entropic pressure of the environment. The decreasing entropy results in rapid, seemingly arbitrary evolutionary innovations, which introduces new species and the population increases. (B) The equal balance of quantum and manifold energies determines Euclidean TF, and low entropy. Interactions form temporal gravity, and deviations of the TF curvature increase entropy. (C) By the end of the era, species that occupy the layers with greatest TF curvature are stripped of living space, precipitating their extinction. The evolutionary era thus ends. The temporal weak nuclear force starts the cycle again on a higher energy level, corresponding to greater morphological and biological complexity.

In the frustrated ecosystem, recovery is slow and complicated by environmental changes, increasing disorder. The increasing entropy spreads genetic elements and potentials for new functions within the ecosystem (such as horizontal gene transfer), leading to the formation of new species. The evolutionary potential is inversely proportional to field curvature, which means that species occupying the central, greatest-curvature layers are stressed most, whereas temporal excess enriches the layers of the ecosystem with lesser curvature. In ecosystems, the species constricted in the layer with the greatest TF curvature (species with inflexible genetic organization) are stripped of their living space (Figures 3.2 and 3.3, region C). The increasing field strength produces desperate, unpredictable behavior which jeopardizes reproduction and leads to extinction. For example, at the end the Cretaceous Period the demise of dinosaurs was not arbitrary or accidental, but the consequence of their inflexible genomic makeup, their inability to change.

The idea of such a multistage evolutionary process is well supported. Important pieces of evolutionary innovations appear well ahead of their evolutionary importance. For example, a substantial part of the molecular architecture necessary for the evolution of the nervous or muscle system evolved in advance (Morris, 2010). Moreover, mutation frequency has been shown to be related to population number! Szendro and colleagues (2013) have examined the entropic trajectory of mutation frequency as a function of population. In contrast to the common expectation of increasingly deterministic evolution, during periods with low genetic concentration, entropy initially decreases and subsequently changes in parallel with the increase in population. This is congruent to the U-shaped curve of Figure 3.3. But the decreasing entropy during the initial stage of evolution is not limited to the laboratory! It is also true for the natural environment and ecosystems. For species ranging from plants to vertebrates, the emergence of new species appears to occur within two million years of an environmental or evolutionary change (Hedges et al., 2015). However, as TF enters the low entropy stage, the species genetic makeup stabilizes.

Other studies that have examined the social aspects of biological systems also support the idea of multistage evolution. During the first vibrant, energetic stage of evolution, species from bacteria and fish to humans appear to lean toward generosity, leading to cooperation and altruism. The generosity appears most prevalent when mutations occur at an appreciable rate, which is only true for the first stage of evolution! This idea was highlighted by evolutionary studies that used the well-known prisoner's dilemma (Stewart et al., 2013). However, when defections sweep through a population, a tipping point is reached and generosity disappears. In the third stage of evolution, cheating becomes the only feasible choice. These findings show the field's control over individual behavior, as recognized by Harsanyi's pioneering work (1968). The above findings defy any other explanation, but can arise as the natural consequence of energetic changes by the temporal weak nuclear force.

The evolutionary process is described here for animal systems, but the same should hold for all eukaryotes. Prokaryotes exist as a large, common gene pool, where environmental conditions are quickly adapted to by horizontal gene transfer, canceling the chance of evolutionary progression. Today, many species of wild animals are perfectly adapted to their living environments and are locked within a great temporal curvature. As a result, they have small genetic diversity and their survival is easily challenged by the degradation of the environment. Although human genetic diversity (evolutionary potential) is low due to historical bottlenecks, such as environmental changes, diseases, and wars, today it appears to be increasing, because we came to occupy the field with the smallest temporal curvature within the ecosystem.

Summary

Within an evolutionary period, the nutrients, energy, and resource cycles must remain balanced. By enforcing their energy neutrality, ecosystems show the characteristics of the Calabi–Yau torus. When the recycling of nutrients is not balanced, the evolutionary period ends and gives way to a new cycle. The first stage of an evolutionary period begins with a small number of surviving species in a renewing environment. The negative TF curvature initiates genetic innovations and structural reorganizations. The abundance of nutrients encourages the formation of predator-prey cycles. The second stage of an evolutionary period occurs within Euclidean TF and is characterized by low entropy. The population expands and competition among the species leads to temporal field curvature differences (life prospect of the species). In the third stage of an evolutionary period, the genetic landscape spreads. The recycling of the nutrients becomes unbalanced, creating stress and frustration. Species occupying the temporal layers with greatest curvature are stripped of living space and die out. In laboratory experiments, the entropic trajectory of mutation frequency versus increasing population decreases and then increases, forming a U-shaped curve. The results provide strong support for the idea of multistaged evolutionary periods. The discrete nature of evolutionary progression is also supported by the appearance of evolutionary innovations well ahead of their evolutionary utility.

Evolution of animals

It must have taken at least a billion years before organic sacks could independently spread, consume, and live. Life emerged as chemical reactions gradually achieved biologically potent, congruent organization around heritable material (the manifold). Once life took hold, it could not be extinguished; entropic pressure and temporal gravity drove it toward more complexity.

The formation of life had a reverberating and irreversible effect on the environment. The temperature, chemical, and physical conditions of the watery Earth changed together with the emerging protolife. As life became more resilient, Earth also became better able to support life. Primitive microorganisms systematically modified the atmosphere by their metabolic processes, by which methane and ammonia were gradually replaced with oxygen and nitrogen. Cyanobacteria fixed the unending solar energy bombarding the Earth. In photosynthesis, cyclical chemical reactions converted carbon dioxide into an energy-rich compound: sugar. Photosynthesis trapped the abundant energy of the sun and inundated the atmosphere with oxygen, having a twofold effect. First, the production of organic material (a highly organized chemical structure) decreased the entropy of the living organisms. Second, the free oxygen increased the entropy of the nonliving environment through oxidation (thus increasing its entropic pressure), which have spurred the evolution of biological innovations. The oxygen-rich atmosphere and the presence of constantly renewing, energy-rich plant matter set the stage for life forms with high energy requirements. Animals used the chemical building blocks and energy of vegetation and the oxidative potential of oxygen for their life processes. The increasing concentration of free oxygen in the atmosphere also produced an increasingly robust ozone layer, which tempered the radiation that reached the surface of Earth. Since multicellular aerobic organisms require more energy than anaerobic organisms, these parallel changes precipitated the fast evolution of multicellularity. On the one hand, sufficient oxygen could be absorbed through the skin (i.e., the outer membrane) for life processes. On the other hand, the abating radiation exposure decreased the mutation frequency, leading to genetic stability and multicellular programming.

The nucleate eukaryotes are divided into the plant and animal kingdoms. Although photosynthesis and plant life are thought to have appeared much earlier than animals, their parallel evolution is in fact much more likely. For example, simple organisms certainly consumed organic materials as soon as organic materials were available to consume. Consumption by animals depends on sensory abilities and digestion.

Movement enables animals to find food and aids them in maintaining homeostasis, which is their low-entropy state. But motility requires an immense amount of energy and necessitates separate digestive system for the absorption and conversion of chemical energy. More evolved organisms require a narrow homeostatic balance for optimal functioning, but their superior sensory abilities and emotional sophistication enhances their ability to exploit resources.

The number of regulators and signaling proteins increases steeply with genome size. Thus, the increased regulatory demand sets an upper limit for prokaryotic genome size and, accordingly, disallows greater complexity. For example, in prokaryotes, the entire genome is replicated from a single origin, whereas the increased genome size and complexity in mammals requires the spatio-temporal coordination of thousands of replication origins. Furthermore, this spatio-temporal order of genome replication changes throughout development and cellular differentiation. In human cells, the number of origins that fire each cell cycle was estimated at 50 000 (Casas-Delucchi & Cardoso, 2011). DNA replication cannot be explained at a genetic level alone. Dual encoding of amino acid and regulatory information has been found to be a fundamental feature of genome evolution (Stergachis et al., 2013).

The folded, secondary, and tertiary structures of proteins prevent misfolding or aggregation and highly constrain the surface, which stabilizes both the protein structure and its gene. For this reason, young proteins and young genes tend to change more easily than more established ones (Chen et al., 2012; Lobkovsky et al., 2010; Tóth-Petróczy et al., 2011; Vishnoi et al., 2010). DNA repair mechanisms limit, slow, or inhibit arbitrary mutations, which permits the gradual increase of organizational complexity, a continuously refined adaptation to environmental changes. The complexity of gene regulation has been increasing continuously over evolution and probably will continue to do so in the future (Warnefors & Eyre-Walker, 2011). Today evolution is understood as an increasingly active cellular process, capable of initiating horizontal DNA transfer, interspecific hybridization, or massive genome restructuring (Saphiro, 2011).

In higher organisms, the number of genes routinely multiplies. Ohno's early seminal work (1970) showed how gene duplication can play an important role in evolution by creating redundancy. Because the stability of protein sequences is tied to the genetic code, gene duplication opened the way for mutations that created modified and related functions. The explosion in gene number in eukaryotes shaped cellular complexity, giving rise to organelles with unique functions such as endoplasmic reticulum, endocytic vesicles, and ribosomes.

The evolution of vertebrae coincided with another explosion—when the gene number more than doubled to tens of thousands. The mind-baffling intricacy of cell regulation indicates that the functional complexity of living systems has no limit and will probably continue to evolve in the future.

Summary

Evolution not only enhanced the complexity and resilience of the living world but also made Earth better able to support life. The oxygenation of the atmosphere oxidized metals and organic materials, increasing environmental entropy (the entropic pressure). The emerging plant matter provided for the high energy requirements of animal life and the presence of atmospheric oxygen allowed the evolution of multicellularity. Motility accords animals the means to find ample food to satisfy the immense energy requirements of movement. The DNA repair mechanism greatly diminishes arbitrary mutations, permitting the evolution of increasing functional complexity. Gene duplication encouraged the development of new, related functions. Today, evolution is considered an increasingly active cellular process, capable of initiating massive genetic changes.

The evolution of the mind

In multicellular animals, the internal organizations of the body and the nervous system—the center of homeostasis—are highly interdependent. In the most primitive bilaterians, the nerves are regulated in a linear fashion (see Chapter 2, „Appearance of the nervous system"). Evagination of pallium, a typical morphological event in the forebrain of almost all vertebrates, preserves a central cavity for the limbic brain, around which the cortical structures can form. The limbic activity is relayed toward the evaginated pallium or cortex, forming a memory trace by activating neurons in these areas. The energy requirement of neuronal activation extinguishes the electromagnetic flow, and a flow-reversing response restores the energy-neutral state (see Chapter 2, „Mechanism of mental operation"). This automatic action turns the brain into a self-contained energy-neural unit: the mind. The mind emerged to interact with the external world, but it became a continuously improving mirror of it, making evolution a progression toward better sensory, motor, and mental functions. The mind can render sensory reality because the subconscious (cortex) evolved as the exact, meaningful representation of the environment. The mind compares sensory information to the information history of the subconscious. In this way, full understanding or an educated guess emerges from approximate, partial information. (We do not have to see the whole of anything to know what it looks like.) This method conceptualizes the environment; with increasing focus the resolution expands proportionally, which is to say, infinitely. This is why differences from the expected emerge first in the mind (Quian Quiroga et al., 2014). Brain activation patterns are responsible for coordinated behavior. When all possible forms of motion have a manifold (neuronal activation pattern) representation, automatically coordinating and optimizing any movement becomes possible, which makes synchronized behavior effortless. Bodily or functional changes and behavior are direct consequences of this mental pattern.

In the temporal elementary particle, interaction forms new neuronal connections or strengthens old ones (a memory trace) in the cortex. Memory potential is the formation of experience by the increasing complexity of neuronal connections and their assemblies. As the neuronal patterns stabilize, the energy requirement of activation becomes progressively smaller; therefore emotional involvement declines, until it disappears altogether. An automatic behavior has formed. In this way complicated regulation (i.e., learning) can form very quickly in mammals. The evolution of the cortex is the perfection of the Calabi–Yau manifold; and mental evolution is the history of corticalization. The ability for nuanced interaction, such as the capacity for preconceived ideas, explodes thanks to emotion. Animals with a limbic brain cannot form emotions, so they are incapable of imposing a conscious threat or of accumulating experience through learning. In mammals, feelings produce hormonal, bodily, and behavioral changes and determine a state of consciousness, which then identifies itself with those emotions (Guterstam, 2015). Feelings extend the self as a tool of manipulation. For example, emotional displays can be just as effective as movement itself to achieve conscious aims. The lion does not have to get up to chase annoying monkeys away from the neighborhood, his roar is sufficient. Emotional hysteresis (getting stressed, angry) is the privilege of emotional beings (see section in Chapter 2 with the same title). The resulting vigilance leads to far better preparedness in dangerous situations, although anger and argument can be just as tiring, or even more so, than physical effort.

Summary

Sensory stimuli activates neurons in the brain of mammals and birds. The energy requirement of neuronal activation extinguishes the electromagnetic flow and forms a memory trace. The emergent electromagnetic imbalance reverses the energy flow, triggering a response, such as motoric activation. Through interaction with the outside world, the mind develops increasingly better representations of it, which makes coordinated behavior effortless because all possible movement can be triggered by the appropriate, automatic neuronal activation pattern. The evolution of the cortex enables increasingly complicated behavior and learning.

Temporal Elementary Particles and the Laws Governing them

Temporal elementary particles

As the reader will recall from Chapter 2, the three types of temporal elementary particles differ in the energy level and complexity of the torus. For temporal (emotional) elementary particles, the increasing neural organization represents an evolutionary progression toward regulatory complexity.

EMOTIONAL NEUTRINO   Emotional neutrinos are simple organisms with linear neural regulation. The limbic brain perfects an organizational congruence with the TF, so it responds to environmental signals with increasing precision. Behavior is hard wired (i.e., it has a genetic origin) and learning remains rudimentary. Unable to produce emotions, sharks exist almost exclusively as food-acquiring machines. These animals can trigger potent fear in the popular imagination precisely because of their emotionless behavior.

Summary

Emotional neutrinos have uninsulated limbic brains and lack emotions. Behavior is hard wired, but exceptional sensory abilities allow fast and precise response to stimuli.

EMOTIONAL ELECTRONS   Animals with well-formed cerebrums form **emotional electrons**, which populate most regions of Earth. The insulated quantum energy forms the self, or ego, which is the source of cognition and self-awareness. In complex organisms, the preservation of the ego becomes the dominant emotionally supported motivation. Emotions are the tool of survival; with them dangers can be avoided or overcome, and opportunities can be found. Animals with more sophisticated emotions appear later in evolution, and these animals exhibit great evolutionary advantages. The discrete energy changes lead to the Heisenberg uncertainty principle, and the Pauli exclusion principle, a competitive drive leading to territorial needs. Emotions dramatically improve homeostatic regulation, such as the ability to maintain constant temperature. Emotions can form the white hole state of love that allow mating and care of offspring.

Summary

Cerebral animals form an energetic unit which is the ego: the source of cognition and self-awareness. These animals form emotions, tools of homeostatic and behavioral regulation. Emotion enables them to be warm blooded, highly social, and give superior care to their progeny. They obey the Heisenberg uncertainty principle and the Pauli exclusion principle, which leads to competitiveness and territorial needs.

EMOTIONAL QUARK   The highest-energy accumulation produces the third, most advanced group of animals: the emotional quark. Capable of forming all emotions (temporal gravity, temporal electromagnetic force, evolutionary weak nuclear force, and emotional strong nuclear force), only humans have evolved or are evolving toward emotional stability and sophistication, which is required to form harmonic, congruent communities.

Summary

Only emotional quarks possess the emotional strong nuclear force.

Newton's laws and the laws of thermodynamics

Material fermions exhibit classical behavior, which involves temperature and pressure. Likewise, individual quantum uncertainty gives way to societies, which operate through conflicts and interactions, which are the hallmarks of emotional temperature and pressure. Newton's laws and the laws of thermodynamics apply not only to matter but also to societies.

NEWTON'S LAWS   Universal gravitation directs the movement of objects in space, but temporal gravity operates over time and forms our belief system and traditions. A low level of temporal gravity is useful because it provides a stable mental ground and a strong purpose, both of which support goals and achievement. However, excessive temporal gravity becomes an insurmountable mental barrier, causing helplessness rooted in fear and the corresponding immobility. Strong emotional ties to things and people slow change and impede progress.

INERTIA   Mental intentions remain constant without outside emotional force. Temporal quantum energy is constant, and MiDT is stable over time, only to be discharged at the opportune time with the full force of a storm. Anger cannot be undone without an emotional encounter.

NEWTON'S SECOND LAW   The net force on an emotional mind is equal to the rate of change. This law is closely related to free will and makes that discussion poignant.

NEWTON'S THIRD LAW   Every action causes an equal but opposite reaction. Emotional interaction is inherently reciprocal, generating equivalent intentions (over time) in emotional beings. Kindness results in kindness and abuse leads to aggravation. In this way our present actions form our future social environment.

ZEROTH LAW OF THERMODYNAMICS   A thermal, civil equilibrium exists between emotional minds; people adopt the emotional temperature of their environment. The environment determines individual behavior, emotional temperature, and emotional attitude. Many studies show that messy, disorganized environments induce violent tendencies, crime, or phobias (Ramos & Torgler, 2010). Over time, people adopt not only the customs but also the moral and emotional temperament of their environment.

FIRST LAW OF THERMODYNAMICS (Conservation of time)   In biological systems time is conserved, temporal energy can transform into information and back. In interacting systems the sum of MiDT and MaDT (i.e., emotion), remains constant. Without interaction, emotion continues unchanged.

SECOND LAW OF THERMODYNAMICS   The entropy of material systems increases from the Euclidean field toward either lesser or greater spatial curvature. However, the second law expresses the second possibility. In ecosystems and societies, entropy decreases initially, but increases in their later phases. Through interactions, ecosystems and societies form their own temporal gravity field, which are the societal, economic, and cultural differences of society. The weak nuclear force allows the process to be repeated over greater manifold energy. For this reason material systems march toward black holes state by degradation, but complexity increases in living systems and societies.

THIRD LAW OF THERMODYNAMICS   People and societies with greater quantum energy form more erratic, heated emotions. People and societies with moderate quantum energy are calmer, more balanced, and steady. However, societies can never have sufficient entropy to be devoid of emotional friction. Interactions beget emotions. Emotions are part of the human condition and part of society.

Summary

Societies and ecosystems obey Newton's laws and the laws of thermodynamics, which play out over time.

Genetic diversity and population entropy

LOW GENETIC DIVERSITY   Genetic diversity parallels changes in population entropy in naturally occurring populations see „Characteristics of an evolutionary era." In animal populations and in the population of hierarchical societies, population entropy is kept low by fear, by the energy expended by predators, and by environmental threats (wars, plagues, or natural disasters in societies). Low genetic diversity (genetic purity) gives strong genetic potential (biological fitness) to wild animals. Indeed, genetic diversity appears to be low in wild mammalian species (Leffler et al., 2012) see Figure 3.3 (region B). Just as a gas squeezed into the corner of a room represents a low-entropy state from which, theoretically, the gas can be used for work as it spreads out over the entire container, low-entropy populations work tirelessly to keep their isolation and attack those who appear to violate their perceived territory. In animals, this manifests itself as territorial protection and violence; in societies this shows up as racist, homophobic, and chauvinistic fervor, which can turn violent. The gazelle that is different will be rapidly hunted down by predators. For example, in an experiment, wild antelopes that were painted white invariable disappeared within a day. Nevertheless, I propose that animals that form emotions are extremely particular about their mating choices and form complex courting rituals, which exclude differing individuals from the mating pool. In turn, careful mating practices ensure genetic purity. In societies, stress causes nonessential activities, such as art making to become simplified, or neglected altogether. In short: stress and art do not mix! Populations with greatly reduced size are especially vulnerable for environmental shocks, which can make the community unfriendly and suspicious, a condition that blocks creativity, as seen in „Mental expansion" of Chapter 2. Archeological records indeed show a disappearance of complex behavior when the population size is significantly reduced (Kline and Boyd, 2010). In case of complete isolation (i.e., a lack of outside influence) the entropy would increase as dictated by the second law of thermodynamics, and as seen on the Galapagos Island.

Summary

Genetic diversity and population entropy change in concert. In wild populations and hierarchical societies, the population entropy is kept low by constant threat (predators, natural disasters, or wars). Just as for gases, low-entropy conditions contain expendable energy that can be used for territorial protection or, in societies, for chauvinism. The resulting apprehension leads to careful mating practices, which in turn maintains the low (genetic) entropy of the population. Population entropy forms the evolutionary potential of the species (genetic fitness) through the regulation of mating.

HIGH GENETIC DIVERSITY   As discussed earlier, both pure quantum energy and pure manifold energy corresponds to high mental entropy and lead to high genetic diversity. High genetic diversity is typical at the start of the evolutionary period, when remnants of the previous ecosystem form a highly diverse, but small population. Plentiful and pleasant environmental conditions create a relaxed, trusting atmosphere that encourages caring and playfulness. As seen in „Stages of evolutionary periods," the high entropy conditions of negative TF curvature encourage generosity across all species (Stewart et al., 2013). People in agreement copy each other's mannerisms and even posture (Cook et al., 2012). Thus, in peaceful, trusting environments copying becomes the dominant way in which new skills are acquired.

Copying occurs without conscious focus or mental effort, so emotional energies enhance social cohesion and trust. Conversely, the inevitable conclusion of an experiment examining human social learning (McNally et al., 2010) is that trust allows copying in the first place. Trust in others, even strangers (e.g., traffic, internet, or product labels) is the nature of modern life which enhances creativity and it is a key toward faster progress.

Today, both genetic diversity and societal entropy are increasing due to democratization on an international scale that has no historical precedent. Such an environment stimulates positive social changes and great scientific advances. New and novel ideas and technological innovations are appearing at an ever-faster pace, and their widespread availability is greater than ever. This flexible milieu aids mental and epigenetic flexibility and complexity, facilitating further social changes. Evolution is not an arbitrary process but leads toward emotional stability. Due to technological and societal progress and change, *Homo sapiens* is the only animal that not only adapts to its changing environment but is also the initiator of profound change. The dynamic circumstances of man make him the fastest evolving (perhaps the only evolving) animal on the planet (Powell, 2011), and the future of humanity will be shaped by individuals (and nations) that occupy a smaller temporal curvature.

Summary

Plentiful, pleasant conditions increase mental entropy, leading to trust and playfulness. Democratization increases population entropy and genetic diversity, which positions humans well for evolutionary change. Humankind is the fastest evolving animal on the planet.

The beginnings of society

The past ten thousand years have seen the evolution of society. Ecosystems and societies evolve by the evolutionary weak nuclear force. However, societal memory is dependent on religion and cultural formations. Without cultural institutions (the societal manifold), such as religion, beliefs, theatre, music, or arts, the history of the chaotic and brutal first empires is lost to time.

Anatomically modern humans emerged about 200 000 years ago, and they were dressing themselves shortly thereafter (Toups et al., 2011). About 50 000 years ago archeological finds show a softening facial structure, which is an indication of decreasing testosterone levels (Cieri et al., 2014) and which probably coincided with declining levels of aggression. However, we must entertain the possibility that a more social climate reduced violent conflict, which consequently led to a reduction in testosterone levels. The social distance decreased, allowing greater cooperation, initiating widespread tool making and use.

During the twentieth century, social distance has again dramatically decreased. The full participation of women in society (women entering the workplace and the sexual revolution) precipitated a new feminization. These changes reformulate the dynamics of social interaction, lessening conflicts and leading to greater cooperation and a more democratic and flexible social environment. A more accepting society nurtures trust, which is conducive to social and technological progress.

The more peaceful environment has naturally increased life expectancy. Living beyond the reproductive age in significant numbers created the social cohesion of intergenerational living, with the older members of the community taking over the lion's share of the care for the infirm and the children. This also allowed the elderly to pass on their accumulated life experience to the next generation. The survival probably improved in every age group, leading to extended families and the creation of a type of relaxed, communal environment, where art-making, music, and storytelling could thrive. Such accumulation of knowledge, beliefs, culture, and tradition became the basis of society. Archeological finds indeed suggest a dramatic increase of longevity in modern humans during the upper Paleolithic, which parallels cultural innovations and population expansion (Caspari & Lee, 2004). In tribal communities, the self identifies fully with the community, which is often an extended family structure. Trust in the natural order of things leads to high societal entropy. The formation of society proper turned all that upside down by replacing the trust of the tribal structure with brutality and fear (the high entropy of quantum energy). In other words, to begin a society from scratch, without infrastructure, is a dangerous affair. In the early stages of a society, personal safety is nonexistent and societal membership is maintained by brute force. All over the world, from China to the Americas, the first societies were highly hierarchal and were maintained by unimaginable brutality. Oppressive regimes battled with their own people and with each other. Human sacrifice and signs of horrific punishments were found in ancient China, Egypt, and the Americas.

In these hierarchical societies high quantum energy corresponds to the high social entropy of disorder, fear, wars, and calamities of all types.

Homo sapiens is a social animal. Even the most primitive community provided some sense of safety to its members by reducing dangers from untamed forests, savannas, and predatory animals. As communities, people could isolate themselves to a large extent from the dangers of the natural environment. In addition, societal rules also soothed conflicts between members. Responsibilities were divided into sharply separated classes. The ruling class was at the top, followed by military, priests, and craftsmen. At the bottom were commoners, peasantry, and slaves. Irrespective of the circadian rhythm, artificial requirements increasingly dictated work and life, as the citizenry lost touch with the natural cycles of the environment. Since the hierarchical world view is static, changes are viewed as threats. In hierarchical societies, individual identity is formed around a prescribed role in the ever-more complex existential structure. Choices between right and wrong are simple and predictable. The scaffolding of hierarchical societies is made of the individual egos of its members. The lack of trust of the egoistic mind leads to rigidity and a readiness to be manipulated see, „Low genetic diversity." People in low-entropy societies are willing to commit repulsive, even despicable acts in the name of nation and even god, resulting in unspeakable, perverse punishments and perverse entertainments (e.g., the Roman Coliseum). Societies that rely heavily on physical force are necessarily patriarchal. However, a few thousand years ago another important mental change occurred. Egypt's rulers were still considered as Gods, but in Greece the gods were people. Man could recognize himself in the failings; the follies, and the adventures of the habitants of Olympus. Democratic ideas were predicated and preached by sages of antiquity. As shown in „Characteristics of an evolutionary era" the individual and social mental entropy changes together. The openness of society determines genetic diversity, which is the variation in the gene pool. Like ecosystems, societies form submanifolds and therefore evolve by the evolutionary weak nuclear force. In their first phase, societies start out with vibrant, optimistic energy, which gives way to long-term stability of the second phase. An evolutionary period always ends in disintegration and revolutions, instigated by increasing inequality. The evolutionary weak nuclear force drives social transformation in a stepwise social evolution. Human history is the reduction of social distance. Today's highly connected world has dramatically accelerated the rate of change. Trust, congruence, and mutual respect allow new ideas and new technologies to spread and accelerate positive change around the globe.

Summary

Societies create some form of culture, called the social manifold. The first societies lacked this energetic support, which made them chaotic and brutal. Hierarchical societies are patriarchic and maintain a rigid social structure. However, societal evolution is a long and slow road toward democratization in which the chaotic quantum energy has gradually transformed into manifold energy, such as culture, religion, scientific understanding, and supporting infrastructure. The analog energetic structures of ecosystems and societies mean that their change occurs according to a common evolutionary process. Societal entropy decreases at the beginning of the cycle and increases afterward as the evolutionary potential gradually gives way to social disintegration. The cycle must terminate and give way to another social formation.

The role of entropy in the economy

The temporal weak nuclear force produces evolutionary change in the ecosystem, in society, and finally, in the economy. In industrial societies, evolutionary progress occurs through economic cycles. Over time, economies maintain energy neutrality (Figure 3.3, the horizontal axis should mark the time of the economic cycle). In economies, market forces, the energetic cycle of raw materials, energy sources, infrastructure, and labor must be in equilibrium. Below I show that energy neutrality exists only in the middle of an economic cycle. The beginning (when an excess in manifold energy triggers the cycle) and the end (when the accumulation of quantum energy leads to disorder and economic collapse) of the economic cycle are opposite-polarity energetic mirrors of each other.

Technological advancements, economic opportunities, and individual mental energy conspire to make the first stage of the economic cycle vibrant. The negative temporal curvature of the weak nuclear force jump starts this enthusiastic and fast-changing phase by lowering societal entropy (Figures 3.2, 3.3, region A). Fast innovations give rise to new industries. The number of companies increases and workers are readily absorbed. Progressive changes and decreasing inequality fuel enthusiasm. These positive feelings of generosity are the result of the nuclear weak force, as discussed in the section „Stages of evolutionary periods." The fast emerging opportunities allow quick return on investments. The fervor of the first phase then gives way to the low-entropy second phase.

The negative temporal curvature flattens out and forms the Euclidean TF. Interactions are regulated by the Pauli exclusion principle. Competition to satisfy the increasingly sophisticated costumers and other market interactions form temporal gravity, and the productivity and size of companies diverge (Figure 3.2, region B). Well-established companies continue their growth, but starting a new venture becomes increasingly challenging. The increasing TF strength marks the start of the third phase of the cycle. Competition between companies becomes fierce as they vie for reluctant costumers. The output outgrows the market, and the energy sources, raw materials, and job opportunities all shrink. Excess debt, the worsening social climate disallows generosity (Stewart et al., 2013). Technological innovations are hard pressed to find their way into industrial production because large, established, and inflexible companies dominate the market. People occupying different layers of temporal curvature experience increasing inequality and widening financial disparity. Companies and individuals occupying the layers with greatest temporal curvature (less flexible, less innovative companies) watch their economic opportunities narrow and bottom out, leading to large-scale failures and bankruptcies (see Figure 3.2, region C). The changes become self-perpetuating, moving the system inevitably toward a high-entropy singularity. This singularity condition is often termed chaos although, in truth, quantum energy increasingly dominates. After great losses and immense suffering, the decimated economy starts a new cycle with a greater technical and intellectual capital. The economic submanifold forms anew, but this time on a higher energy level, as economic evolution moves on.

This complex process is supported by recent studies (Greenhouse, 2008; Piketty, 2014; Compton, 2004). Piketty poured over 200 years of tax records in several countries to reach the conclusion that, in developing societies, inequality is muted and in maturing societies inequality is increasing. During the evolution of ecosystems, the species with high temporal mass becomes extinct; in economies, inflexible companies go bankrupt. Indeed, economic progress appears to be cyclical, following a stepwise progression that is punctuated by recessions and depressions, such as the long depression in Britain and the United States from 1873 to 1896, or the Great Depression, which affected most national economies of the world from 1929 to 1933. The above argument, which shows that inequality cannot continue to grow unabated, makes Piketty's message on inequality especially timely and poignant.

Summary

Economies maintain energy neutrality over time. The negative temporal curvature (manifold-energy excess) makes the first stage of the economic cycle vibrant. Fast innovations give rise to new industries, new companies, and quick returns on investments. The second stage is characterized by Euclidean curvature (true energy neutrality), the size and success of the companies begins to deviate due to competition (the Pauli exclusion principle). In the third stage, the TF curvature is positive (quantum energy excess). Increasing inequality and widening financial disparity push some people into poverty while the top earners see their financial fortunes boom. The destabilization of energy sources, raw materials, and labor cycles leads to crisis, so the cycle must terminate. After a chaotic phase, the society recovers and reformulates at a greater technological, social, and economic complexity.

The Universe Fine-tunes its Parameters by Forming a Minimal Surface

On its largest scales, the universe has a distinct, well-recognizable cellular structure (Alfvén, 1981; Weygaert, 2007; Cantalupo et al., 2014), which is regulated by two opposing effects. The overall spatial structure of the universe forms smooth manifold, which degenerates close to its poles and forms singularities. The degrees of freedom increase in white holes and decrease in black holes. Thus, both the spatial and temporal surfaces are oriented between and bounded by the poles (Figures 1.4 and **3.4**). However, the microdimensions form entanglement; therefore maintaining a closed and minimal surface (**Figure 3.5** and 3.1, concentric circles). The spatial expansion is kept at bay by the enormous field strength of the black holes (Figure 3.1). However, the constant generation of energy in white holes can reduce the overall curvature of space and lead to Einstein's lambda (i.e., expansion).

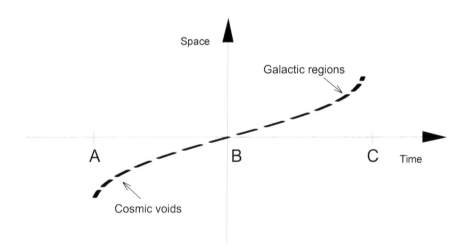

Figure 3.4. Possible energy changes in the universe Capital letters indicate energetic regions: white hole (A), Euclidean region (B) and black hole (C). The horizontal axis represents time (going from negative to positive time) and the vertical axis is space (going from negative to positive microdimensional volume). The spatial curvature changes from positive near black holes to negative near white holes. The changing field curvature gives rise to a power-law distribution. The gravitational entanglement, occurring orthogonally to the field curvature, produces a bell distribution.

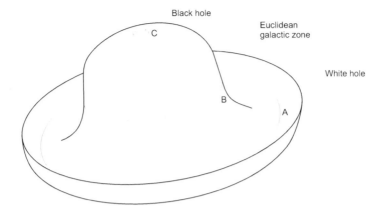

Figure 3.5. Changes in field curvature of the universe The changes in spatial (temporal) curvature between the white holes and the black holes are shown. The region (C) with positive curvature culminates in a black hole. The Euclidean-curvature zone (B) is represented by the middle, noncurving region of the hat, whereas the region (A) with negative curvature is represented by the brim of the hat. The concentric curves (A), (B), (C) show regions of constant field curvature formed by the microdimensions.

By extending the Gerard t'Hooft assumption that all elementary particles are black holes, all elementary-particle boundaries are expected to contain their information history (Susskind, 1994). Lorentz transformations between macro- and microdimensions form inverse volume changes, which enhance field curvature and pressure differences, forming the spatial-temporal complexities of the gravitational field (Figure 1.7.d). Although individual tori can develop notable differences in energy potential, the neighboring field strength tends to equalize. As a result, the universe forms smooth spatial and temporal topology, as the frequencies of standing waves forms gradual succession between the black and white holes. Calculating the SF curvature of Earth's surface (the exact topological distance from black holes or from white holes) should be possible.

Summary

The poles (black holes and white holes) form the forbidding boundaries of the SF and TF. But the microdimensions form closed minimal-surface foam with constant extrinsic curvature. These opposing effects turn the universe into a self-regulating system. During interaction, space and time interchange by Lorentz transformation. In down spin decoherence the field strength and pressure increases, and information accumulates. Up spin decoherence is the opposite process. These contrasting regulatory effects give rise to the complex structure of the cosmos.

BELL-CURVATURE AND POWER-LAW DISTRIBUTION   The occurrence of the Calabi–Yau space on three different levels of the universe gives rise to temporal and spatial submanifolds on many levels. Entanglement conserves the curvature of the field, but the curving field has directionality, which directs the decision of the participant, and the choice of the participant further modifies the field (see Figures 3.4 and **3.6**.b). Within a curving field interaction does not constitute entanglement, because the field is deterministic, allowing only one outcome (Figures 2.3.a and 2.3.d). As discussed in „The anatomy of interaction" in Chapter 1, the changes within the curving field become self-perpetuating and irreversible. As a result, the center of the field empties out, leading to a power-law distribution (Krioukov et al., 2012). Examples of power-law distributions are the formation of internet traffic, the size of Earthquakes, or neural activation patterns.

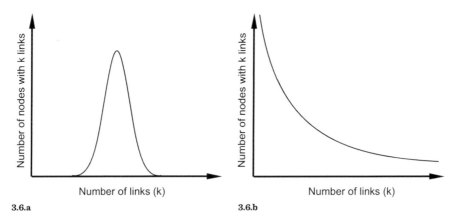

3.6.a   3.6.b

Figure 3.6. Bell-curve distribution and power-law distribution (a) The bell distribution occurs due to gravitational interactions. (b) The power-law distribution is the consequence of the curving field and, inversely, these interactions enhance the curvature of the field.

Entanglement produces two polar end products or particles, one of which is the observer. Thus the observer is part of the system and changes together with the system (Figures 2.3.b,c). Entanglement ensures energy conservation during the partitioning of a common wave function and gives rise to minimal surface foam. It is an orthogonal relationship, which thus conserves the field's curvature (see Figures 3.4 and 3.6.a). Entanglement is a local interaction, which nevertheless moves energies and corresponding field curvature over immense temporal or spatial distances of the TF or SF, forming the topological surface of the cosmos. The analog operation between gravity and harmonic motion was discussed in the section „Relationship between spring motion and gravity" in Chapter 1. The underlying field strength determines the energy distribution between the two entangled parties (how skinny or wide is the shape of the curve). The entropy produced by the gravitational relationships in wide-ranging fields, and even in those connected to biological systems, societies, and economies, has been corroborated by numerous studies (Kleidon, 2010; Krioukov et al., 2012; Wissner-Gross & Freer, 2013; Nathaniel, 2011). Accordingly, Chester (2012) proposed a second-order differential equation of harmonic motion to describe population dynamics. The frequency of harmonic motion corresponds to the event frequency in gravitational relationships, and the field curvature is analogous to the tension of the spring. In this way, the equation for harmonic motion can describe the many forms of gravitational relationships in the material world, in biological systems, or in society.

The opposite polarities of mental expansion and mental contraction as well as love and depression were presented in Chapter 2. Mental expansion, which requires acceptance and equanimity, is literally fueled by the expansionary energy of the universe. It is an energetic state bestowed on individuals with a positive attitude, an open mind. The polar states of mental expansion and aggravation can be considered entanglements, which occur orthogonally to the temporal field, whereas love and depression can be considered white and black holes states occurring parallel with the temporal field. Ultimately death occurs due to resignation, an emotional weakness, and can be viewed a polar opposite of the newborn state (love).

Summary

Interaction within the universe can occur either orthogonally to the space-time axis or parallel to it. Orthogonal interactions are gravitational entanglements, forming bell-curve distributions. Parallel interactions are influenced by the underlying field curvature and lead to power-law distributions.

Conclusions

Evolution is the overarching organizing principle of the universe.

As the Big Bang gives birth to material particles, evolution begets temporal elementary particles. Evolution appears to be a random process, but over time it forms an arch that spans the formation of material fermions and the emergence of the mind. The Pauli exclusion principle increases the differences in field curvature (spatial or temporal) of the universe, creating its poles, which form its boundaries. But the microdimensions (both time and space) form a closed minimal surface through entanglement. Their opposing dynamics lead to self-regulation, which is a continuous fine-tuning of the physical parameters of the universe.

In the vastness of the cosmos, immense differences in temperature, pressure, and entropy produce physical, chemical, biological, and societal evolution. The subsequent evolutionary phases are built on the evolutionary potential of the previous era. Membranous organisms increase order by taking advantage of the entropy increase of their environment. The temporal weak nuclear force drives biological evolution and gives rise to submanifolds of evolutionary periods and eras. Evolutionary periods are separated by unidirectional barriers that can only be crossed when the genetic, organizational, and morphological innovations have spread sufficiently widely within the ecosystem. When the ecosystem outgrows its boundaries, the energy cycle is upended and the evolutionary period or era ends, giving way to the next one, which will have greater biological complexity based on the genetic potential accumulated during the previous era. The sparse surviving population is subjected to significant entropic pressure by the dynamically changing environment. The system, which in this way positions itself toward the TF, develops an orthogonal manifold (the heritable material) and quantum energy (the protein structure). Thus, the genes form a memory of the organizational changes of proteins. In this way, genes evolve in an orthogonal, interdependent relationship with the protein structure. During evolution, complexity is enhanced according to the same orthogonal relationship: bodily organization goes hand in hand with the development of the brain (manifold) that regulates it. In this way, every evolutionary phase is an increase of dimensions, leading to progressively greater complexity! Evolution of the cortex allows the accumulation of experience and leads to the intelligent mind, which can form societies. The first societies emerge as oppressive, brutal regimes, due to chaotic, overpowering quantum energy. Societal and economical evolution, like all evolutionary processes, occurs in successive phases. In each phase the crude experimentation gives way to considerable sophistication in later stages. Democratic society is the perfection of the social manifold over millennia. Intelligent occupants of the cosmos should be similar not only in the structure of their minds, but in the biological building blocks of life, in outside appearance, and in their emotional sophistication as well! The following technologically feasible experiments or tests can verify or contradict the tenets of the proposed hypothesis:

1. The second law of thermodynamics is expected to be violated in gravity-free space. Entropy changes should reflect the difference in spatial curvature between systems interacting in space as opposed to on Earth.

2. In a controlled experiment on the space station in 2007, significant genetic changes were observed in astro escherichia coli bacteria, which became more virulent, and three times more potent. Tardigrades, small aquatic animals had shorter development times in microgravity. Mammalian aging also speeds up during space flight. The above results might have a common origin and confirm that gravity slows not only material, but biological clocks as well.

3. All mammalian brains operate at the same brain frequencies. The hypothesis purports that the evolutionary stability of mental frequencies requires the ratio of the cortical surface area and the limbic volume to remain constant throughout evolution. The ratio of cortical surface area to brain (or limbic) volume can be measured.

4. Within dynamically changing, high-entropy environments the tendency for order and chemical organization should increase in vesicles (or liposomes or cells stripped of cytoplasm and nucleus). A positive finding would verify the existence of environmental entropic pressure.

5. Computer simulations could verify the model and also provide more detailed information and predictions for a variety of phenomena such as the expansion of the universe, economic cycles, internet traffic, or chaotic systems. For example, to test the hypothesis, a computer simulation of the brain could model sensory operation (as the function of fast and slow brain oscillations).

6. Stimuli cause predictable increases in the frequencies of the brain, but it is not known how brain frequencies change over time. Following the brain frequencies between stimulus and response for extended times (several hours) could uncover how the mind recovers its energy-neutral state.

7. The recognition that systems separated by a manifold are subjected to an entropic force allows us to propose ingenuous ways to harvest this green energy. For example, green, cheap energy production in membranous sacks placed in dynamically changing environments should be possible. In fact, the internet can be considered a manifold-forming system.

8. The existence of the information-insulating horizon (manifold) is currently only a speculation of string theory. However, if the brain, ecosystems, and societies are shown to have the same energetic demarcations as elementary particles, then string theory can be experimentally tested and studied.

Undeniably, the proposed hypothesis is in its infancy. It is an ambitious, overarching idea, encompassing theoretical physics, neurology, and evolutionary biology. Glitches, inconsistencies, and other imperfections in its formulation may be ironed out by careful attention from experts in their respective fields.

Afterword

Darwin's idea of evolution was so shocking that he waited twenty years to present it to the wide community. Eventually his idea became the basis of objective investigation of changes in the living world. The hypothesis detailed here posits unexpected connections but, it is supported by a long list of well-respected studies and experiments. The following are the main consequences of the hypothesis:

– The universe has a discrete structure, which leads to the Heisenberg uncertainty principle.

– The Calabi–Yau space formed during the Big Bang and gave rise to space and time.

– The speed of light in vacuum is the speed limit of information transfer.

– The proposed hypothesis opens the way toward the unification of string theory, quantum mechanics, and general relativity.

– The operation of the universe (such as accelerating expansion and excess matter) is explained without invoking dark matter or dark energy.

– Gravity is the consequence of the Pauli exclusion principle, and the uninsulated torus leads to electromagnetism.

– The geometry (spatial or temporal) of the weak nuclear force corresponds to its high mass, which determines the immense mass differences between the families of fermion particles.

– The proposed hypothesis defines time as the information memory of the Calabi–Yau manifold and mass as the eccentricity of the quantum energy (the extent of the elongation of the waveform). Weight is the result of local gravitational curvature. Therefore, temporal weight is typical for the relationship.

– Material evolution gives rise to an orthogonal, oriented, but symmetric structure: the temporal fermion.

– For material fermions, quantum waves spread over space, but for the orthogonal temporal elementary particles, quantum waves exist in temporal freedom.

– Biological systems, society, and economies are also governed by the same physical laws as govern matter. However, the mind's control over the body increases its degrees of freedom.

– The proposed hypothesis explains emotions and classifies them. The elementary forces of the mind are the emotional equivalent of gravity, electromagnetism, and the weak and strong nuclear forces.

– Space, which is energy and time (the latter of which is information) can transform into each other through interaction. Now we can even recreate „spacetime" as the result of two complex dimensions.

– Evolution is the inherent organizing potential of the universe and necessarily converges toward complexity and the intelligent mind (the emotional quark) wherever the necessary minimal conditions exist.

This is the first hypothesis that gives a physical basis for motivation and human behavior. It is also the first overarching idea that encompasses physics, the mind, and evolution. The universe is made up of orthogonal spatial and temporal energies. The existence of the Calabi–Yau space on three different levels of the universe (the material fermion, the emotional fermion, and the universe itself) would explain how the submanifolds in the material world and biological systems and society would form similar phenomena, such as power-law distributions in geology, physics, population dynamics, biology, sociology, and other fields (Koonin, 2011). When the manifold and quantum energies are in relative balance with each other, entropy is low and constant interactions actually keep the field Euclidean. The Pauli exclusion principle leads to entanglement, a bell-shaped distribution, and complexity.

This idea suggests a possible organizational principle of the universe, which supports the inherent intuition of humanity throughout the ages and cultures that an underlying unity connects all of existence. In congruent with the Aristotelian physical world view, reality is considered an evolving complex organic system, which abhors vacuum. Matter and the universe have symmetric energy structures, but vastly different degrees of freedom. Being positioned between matter and the universe, the mind shows the characteristic of both. The negative attitude mind is fully directed by the environment (making it matter like). Only the positive attitude mind has free will (making it God-like). However, the positive attitude mind has no incentive, or motivation to change. The idea that evolution, including societal evolution, is guided toward complexity and order suggests a shining future for the intelligent occupants of the cosmos, including mankind.

Future Prospects

"We must learn to live together as brothers or perish together as fools."

—Martin Luther King Jr.

The proposed hypothesis introduces a new physical world view that requires a paradigm shift in the sciences. If proven correct, the ideas detailed here will open enormous opportunities in a variety of field, such as physics, neurosciences, social sciences, quantum computing, and evolution. The behavior of elementary particles could very well originate in the Calabi–Yau torus, where changes of standing-wave frequencies would free energy in quanta, called elementary forces. An energy-neutral system could form the basis of a quantum computer, which could evolve and would not require special conditions (such as low temperature) for its operation. A deeper understanding of evolution would permit us to exploit it for the production of food or medicine and would lead to a revolution in conservation and environmental protection. Taking advantage of the entropic pressure of Earth's gravity would allow us to design low-cost, safe, and effective energy-producing systems. Studying the mind as a physical system and an elementary fermion can open new avenues in social sciences and psychology. Brain stimulation is already gaining medical importance. Understanding consciousness as the energetic function of a cortical brain would greatly facilitate the efficacy of these methods, opening a path for clinical treatments of a host of brain conditions from schizophrenia to depression and dementia. Today, so many people live in emergency survival mode. Unable to even contemplate their existence and understand the meaning of their lives, their human potential is wasted. Martin Luther King's words never carried more meaning than they do today. We live in a small world. A new idea can multiply and grow in a matter of hours instead of the months and years it took even twenty years ago. Environmental and climate problems urge concerted action by the global community. A disease cannot be isolated even in the backwaters of Africa, and emotional pains can resonate halfway around the globe. Success and creativity is a human right and should be available to every man, woman, and child. Understanding how the mind works would allow us to reach our human potential and feel the joy that comes with achievement. When everyone feels confident of his or her mental power, only then will terrorism and crime become useless and unnecessary for achieving personal or religious aims.

References

Aarts, H.and Van Den Bos, K. (2011). On the foundations of beliefs in free will: intentional binding and unconscious priming of self-agency. *Psychological Science, 22* (4): 532–537.

Ade, P. A. R., Aghanim, N., Armitage-Caplan, C., Arnaud, M., Ashdown, M., Zonca, A. (2014). Planck Collaboration 2013 results. XVI. Cosmological parameters. *Astron.* Astrophys. 571.

Ainsworth, M., Lee, S., Cunningham, M. O., Traub, R. D., Kopell, N. J., and Whittington, M. A. (2012). Rates and rhythms: a synergistic view of frequency and temporal coding in neuronal networks. *Neuron, 75* (4): 572–583.

Aknin LB, Hamlin JK, and Dunn EW (2012). Giving Leads to Happiness in Young Children. *PLoS ONE, 7* (6): e39211.

Alfvén, H. (1981). Cellular Structure of Space. *Cosmic plasma*, Dordrecht. Section VI.13.1.

Almheiri, A., Marolf, D., Polchinski, J., Sully, J. (2015). Black holes: Complementary or Firewalls? *ArXiv: 1207.* 3123v4 {hep-th}

Anderson, C. A., Shibuya, A., Ihori, N., Swing, E. L., Bushman, B. J., Sakamoto, A., Rothstein, H. R., Saleem, M., and Barlett, C. P. (2010). Violent video game effects on aggression, empathy, and prosocial behavior in Eastern and Western countries: A meta-analytic review. *Psychological Bulletin, 136* (2): 151–173.

Arsenault, J.T., Rima, S., Stemmann, H., and Vanduffel W. (2014). Role of the Primate Ventral Tegmental Area in Reinforcement and Motivation. *Current Biology,* doi.org/10.1016/j.cub. 2014.04.044.

Barbey, A. K., Colom, R., and Grafman, J. (2014). Distributed neural system for emotional intelligence revealed by lesion mapping. *Social Cognitive and Affective Neuroscience, 9* (3): 265–272.

Basbaum AI, and Fields HL (November 1978). Endogenous pain control mechanisms: review and hypothesis. *Ann. Neurol., 4* (5): 451–462.

Beggs, J. M., & Timme, N. (2012). Being Critical of Criticality in the Brain. *Frontiers in Physiology,* doi:10.3389/fphys.2012.00163.

Bell, J. S., (1987). Speakable and Unspeakable in Quantum Mechanics, Cambridge: Cambridge University Press.

Bengson, J. J., Kelley, T. A., Zhang, X., Wang, J.-L., and Mangun, G. R. (2014). Spontaneous Neural Fluctuations Predict Decisions to Attend. J *Cog. Neuroscience.,* doi:10.1162/jocn_a_00650.

Bethell EJ, Holmes A, MacLarnon A, and Semple S (2012). Evidence That Emotion Mediates Social Attention in Rhesus Macaques. PLoS ONE7(8): e44387.

Bérut, A., Arakelyan, A., Petrosyan, A., Ciliberto, S., Dillenschneider, R., and Lutz, E. (2012). Experimental verification of Landauer's principle linking information and thermodynamics. *Nature (London) 483:* 187–190.

Blaylock, G. (2010). The EPR paradox, Bell's inequality, and the question of locality. *American J. Phys. 78* (1): 111–120.

Bliss, C. A., Kloumann, I. M., Harris, K. D., Danforth, C. M., and Dodds, P. S. (2012). Twitter reciprocal reply networks exhibit assortativity with respect to happiness. arXiv:1112.1010: 1–10.

Bliss, T, and Lomo, T. (1973). Long-lasting potentiation of synaptic transmission in the dentate area of the anaesthetized rabbit following stimulation of the perforant path. *Journal of Physiology,* 232: 331–356.

Bohm, D., (1952). A Suggested Interpretation of the Quantum Theory in Terms of 'Hidden' Variables. I and II, Physical Review, 85: 166–193.

Bordács, S., Kézsmárki, I., Szaller, D., Demkó, L., Kida, Murakawa, N. H., and Onose, Y. (2012). Chirality of matter shows up via spin excitations. *Nat. Phys., 8,* 734–738.

Bovaird, T., and Lineweaver, C. H. (2013). Exoplanet predictions based on the generalized titius-bode relation. *Monthly Notices of the Royal Astronomical Society, 435* (2): 1126–1138.

Brady, R., and Anderson, R. (2014). Why bouncing droplets are a pretty good model of quantum mechanics. *arXiv:*1401.4356.

Brembs, B. (2011). Towards a scientific concept of free will as a biological trait: spontaneous actions and decision-making in invertebrates. Proceedings.Biological sciences / The Royal Society, 278(1707), 930–939. doi:10.1098/rspb.2010.2325.

Brissaud, J-B. (2005). The meanings of entropy. *Entropy, 7* (1): 68–96.

Brunner, N., and Linden, N. (2013). Connection between Bell nonlocality and Bayesian game theory. *Nat. Commun.,* 4: 1–7.

Burgess, C. P. (2007). Lectures on cosmic inflation and its potential stringy realizations. *arXiv: 0708.2865 hep-th*/0610320.

Busemeyer, J. R., and Bruza, P. D. (2012). Quantum Models of cognition and decision, Cambridge University Press, New York.

Candelas, P., Horowitz, G., Strominger, A., Witten, E. (1985). Vacuum configurations for superstrings. *Nucl. Phys., 258:* 46–74,

Buzsaki, Gy (2011). Hippocampus, *Scholarpedia, 6* (10): 1468.

Buzsáki, G., and Silva, F. L. Da. (2012). High frequency oscillations in the intact brain. *Progress in Neurobiology, 98* (3): 241–249.

Buzsáki, G., and Wang, X.-J.(2012). Mechanisms of Gamma Oscillations. *Annual Review of Neuroscience.,* 35: 203-25.

Buzsaki, G., Logothetis, N., and Singer, W. (2013). Scaling brain size, keeping timing: Evolutionary preservation of brain rhythms. *Neuron. 80* (4): 751–764.

Canetti, L., Drewes, M., and Shaposhnikov, M. (2012). Matter and antimatter in the universe. *New J. Phys., 14* (9): 095012.

Cantalupo, S., Arrigoni-Battaia, F., Prochaska, J. X., Hennawi, J. F., and Madau, P. (2014). A cosmic web filament revealed in Lyman-α emission around a luminous high-redshift quasar. *Nature (London), 506* (7486): 63–66.

Capek, V., and Sheehan, D. P. (2005). Challenges to the second law of thermodynamics (Springer, Netherlands).

Carrick, J., Turnbull, S. J., Lavaux, G., and Hudson. M. J. (2015). Cosmological parameters from the comparison of peculiar velocities with predictions from the 2M density field. Monthly Notices of the Royal Astronomical Society, 450 (1): 317 DOI: 10.1093/mnras/stv547

Casas-Delucchi, C.S., and Cardoso, M.C. (2011). Epigenetic control of DNA replication dynamics in mammals. *Nucleus., 2* (5): 370–82.

Caspari, R., and Lee, S. (2004). Older age becomes common late in human evolution, *PNAS, 101* (30): 28–33.

Challinor, A. (2012). CMB anisotropy science: a review. arXiv:1210.6008v1,288, 1–11.

Chen, W.-H., Trachana, K., Lercher, M. J., and Bork, P. (2012). Younger genes are less likely to be essential than older genes, and duplicates are less likely to be essential than singletons of the same age. *Molecular biology and evolution, 29* (7): 1703–6.

Chen, F.S., Minson, J.A., Schöne, M., and Heinrichs, M. (2013). In the eye of the beholder: Eye contact increases resistance to persuasion. *Psychological Science, 24* (11): 2254–2261.

Chester, M. (2012). A Fundamental Principle Governing Populations. *Acta biotheoretica,* doi: 10.1007/s10441-012-9160-6.

Christoff, K., Gordon, A. M., Smallwood, J., Smith, R., and Schooler, J. W. (2009). Experience sampling during fMRI reveals default mode network and executive system contributions to mind wandering. *Proceedings of the National Academy of Sciences of the United States of America, 106* (21): 8719–8724.

Cieri, R. L., Churchill, S.E., Franciscus, R.G., Tan, J., and Hare, B. (2014). Craniofacial feminization, social tolerance and the origins of behavioral modernity. *Current Anthropology, 55,* 4: 419–443.

Cline, J. M. (n.d.). (2014). STRING COSMOLOGY. *arXiv:* hep-th/0612129v4, 1–47.

Compton, C. (2004). All Men Created Unequal: Trends and Factors of Inequality in the United States. *Issues in Political Economy, 13,* 1013–1023.

Cook, R., Bird, G., Luenser, G., Huck, S., and Heyes, C. (2012). Automatic imitation in a strategic context: Players of Rock-Paper-Scissors imitate opponents' gestures. *Proceedings of the Royal Society B: Biological Sciences, 279,* 780–786.

Csikszentmihalyi, M., & Hunter, J. (2003). Happiness in everyday life: The uses of. *Journal of Happiness Studies, 4* (January): 185–199.

Custers, R., and Aarts,H. (2010). The Unconscious Will: How the Pursuit of Goals Operates Outside of Conscious Awareness. *Science, 329* (5987): 47–50.

Day MV, and Bobocel DR (2013). The Weight of a Guilty Conscience: Subjective Body Weight as an Embodiment of Guilt. *PLoS ONE, 8* (7): e69546.

Danner, D. D., D. A. Snowdon and W. V. Friesen. (2001). Positive emotions in early life and longevity: Findings from the nun study. *Journal of Personality and Social Psychology, 80:* 804–813.

Deaner, R. O., Isler, K., Burkart, J., & Van Schaik, C. (2007). Overall brain size, and not encephalization quotient, best predicts cognitive ability across non-human primates. *Brain, Behavior and Evolution, 70* (2): 115–124.

Dieks, D., & Lubberdink, A. (2011). How Classical Particles Emerge From the Quantum World. *Found. Phys., 41* (6): 1051–1064.

Diener, E., & Chan, M. Y. (2011). Happy People Live Longer: Subjective Well-Being Contributes to Health and Longevity. *Applied Psychology: Health and Well-Being, 3* (1), 1–43.

Dixon, M. L. (2010). Uncovering the neural basis of resisting immediate gratification while pursuing long-term goals. *The Journal of Neuroscience, 30*(18): 6178–6179.

Dmochowski, J. P., Bezdek, M. a., Abelson, B. P., Johnson, J. S., Schumacher, E. H., and Parra, L. C. (2014). Audience preferences are predicted by temporal reliability of neural processing. *Nat. Commun., 5:* 1–9.

Dolan RJ. (2008). Neuroimaging of cognition – past, present and future. *Neuron, 60:* 496–502.

Duncan, J. F. R., Griffin, M. J., and Ono, K. (2015). Proof of the Umbral Moonshine Conjecture. *arXiv: 1503.* 01472v1 [math . RT]

Duncan, T. L., and Semura, J. S. (2004). The Deep Physics Behind the Second Law : Information and Energy As Independent Forms of Bookkeeping. *Entropy, 6* (1): 21–29.

Dudek, S, Bear, MF. (1992). Homosynaptic long-term depression in area CA1 of hippocampus and effects of N-methyl-D-aspartate receptor blockade. *Proceedings of the National Academy of Sciences, 89:* 463–467.

Dupret D, O'Neill J., and Csicsvari J. (2013). Dynamic Reconfiguration of Hippocampal Interneuron Circuits during Spatial Learning. *Neuron, 78:* 166–180.

Dreber, A., Gerdes, C., and Gransmark, P. (2013). Beauty queens and battling knights: Risk taking and attractiveness in chess. *Journal of Economic Behavior and Organization, 90*: 1–18.

Ehrenfreund, P., Spaans, M. and Holm, N. G. (2011). The evolution of organic matter in space. Phil. Trans. R. Soc. doi:10.1098/rsta. 2010.0231.1936538–554.

Eimer, M. (1999). Facilitatory and inhibitory effects of masked prime stimuli on motor activation and behavioural performance. *Acta Psychologica, 101*: 293–313.

Engel, J., and da Silva, F. L. (2012). High-frequency oscillations—Where we are and where we need to go. *Progress in Neurobiology, 98* (3): 31–68.

Ewing, S. G., and Grace, A. A. (2013). Long-term high frequency deep brain stimulation of the nucleus accumbens drives time-dependent changes in functional connectivity in the rodent limbic system. *Brain Stimulation, 6* (3): 274–285.

Fell, J (2012). I Think, Therefore I Am (Unhappy). *Front. Hum. Neurosci., 6*: 132.

Festinger, L. (1957). The theory of cognitive dissonance, Stanford (CA) Stanford University Press.

Filler, A. (2010). The History, Development and Impact of Computed Imaging in Neurological Diagnosis and Neurosurgery: CT, MRI, and DTI. *Internet Journal of Neurosurgery, 7* (1): 5–81.

Fingelkurts, A. A., and Fingelkurts, A. A. (2015). Operational Architectonics Methodology for EEG Analysis: Theory and Results, *Neuromethods*, Ed. Sakkalis V. Springer 1–58.

Fjell, A. M., Walhovd, K. B., Fischl, B. and Reinvang, I. (2007). Cognitive function, P3a/P3b brain potentials, and cortical thickness in aging. *Hum. Brain Mapp., 28*: 1098–1116.

Fredrickson, B. L. and Joiner, T. (2002). Positive emotions trigger upward spirals toward emotional well-being. *American Psychological Society, 13.* 2: 172–175.

Fredrickson, B. (2003). The value of positive emotions: The emerging science of positive psychology is coming to understand why it's good to feel good. *American scientist, 91* (4): 330–335.

Frenk, C. S., and White, S. D. M. (2012). Dark matter and cosmic structure. *Annal. Phys. 524*: 507–534.

Fu, W., O'Connor, T. D., Jun, G., Kang, H. M., Abecasis, G., Leal, S. M., and Akey, J. M. (2013). Analysis of 6,515 exomes reveals the recent origin of most human protein-coding variants. *Nature (London), 493* (7431): 216–20.

Gal, D., and Rucker, D. D. (2010). When in doubt, shout! paradoxical influences of doubt on proselytizing. *Psychological Science : A Journal of the American Psychological Society* / APS, 21(11): 1701–1707.

Gieseler, J., Quidant, R., Dellago, C., and Novotny, L. (2014). Dynamic Relaxation of a Levitated Nanoparticle from a Non-Equilibrium Steady State. *Nat. Nanotechnol.*, AOP, DOI: 10.1038/NNANO.2014.40.

Gilbert, D. T., Killingsworth, M. A., Eyre, R. N., and Wilson, T. D. (2009). The surprising power of neighborly advice. *Science*, New York (NY), 323 (5921): 1617–1619.

Gorno-Tempini, M.L., Rankin, K.P., Woolley, J.D., Rosen, H.J. Phengrasamy, L., and Miller, B.L. (2004). Cognitive and behavioral profile in a case of right anterior temporal lobe neurodegeneration. *Cortex, 40*: 631–644.

Greenhouse, S. (2008). The Big Squeeze: Tough Times for the American Worker. New York (NY) Anchor, Random House LLC.

Greitemeyer, T., and Osswald, S. (2011). Playing prosocial video games increases the accessibility of pro-social thoughts. *J of Personality and Social Psychology, 98*, (2): 211–221.

Grover, L. K. (1997). Quantum mechanics helps in searching for a needle in a haystack. Physical Review Letters, 79: 325-328.

Guterstam, A., Abdulkarim, Z., and Ehrsson, H. H., (2015). Illusory ownership of an invisible body reduces autonomic and subjective social anxiety responses. *Scientific reports. 5*, 9831 doi: 10. 1038/srep09831

Harsanyi, J. C., 1967/1968. Games with incomplete information played by Bayesian Players, I-III. *Management Science, 14* (3): 159–183 (Part I), 14(5): 320–334 (Part II), 14(7): 486–502 (Part III).

Hawking, S. (1988). A Brief History of Time. New York (NY) Bantam Dell Publishing Group.

Hebb, D. O. (1949). The Organization of Behavior: A Neuropsychological Theory. New York: Wiley and Sons.

Hedges, S. B., Marin, J., Suleski, M., Paymer, M., and Kumar, S. (2015). Tree of Life Reveals Clock-Like Speciation and Diversification. *Mol Biol Evol., 32* (4): 835–845.

Hill, A.L. Rand, D.G. Nowak, M.A. and Christakis, N.A. (2010). Emotions as Infectious Diseases in a Large Social Network: The SISa Model, *Proceedings of the Royal Society*, B277(1701): 3827–3835

Holland, R. W., Hendriks, M., and Aarts, H. (2005, a). Smells like clean spirit. *Psychological Science, 16* (9): 689–693.

Holland, R. W., Hendriks, M., and Aarts, H. (2005, b). Smells like clean spirit. Nonconscious effects of scent on cognition and behavior. *Psychological Science, 16* (9): 689–693.

Howard, M. W., Rizzuto, D. S. and Caplan J. B. (2003). Gamma oscillations correlate with working memory load in humans. *Cerebral Cortex. 13*, 12: 1369–1374.

Hudgins, D. M. (2002). Interstellar Polycyclic Aromatic Compounds and Astrophysics. *Polycyclic Aromatic Compounds, 22* (3–4): 469–488.

Kaiser, R. I. Stockton, A. M. Kim, Y. S. Jensen, E. C. and Mathies, R. A. (2013) On the formation of dipeptides in Interstellar Model Ices. *Ap. J., 765* (2): 111–120.

Iacovlenco, O. (2012). An Introduction to Hamilton' s Ricci Flow. Department of Mathematics and Statistics, McGill University, Montreal, Quebec: 1-16.

Ijjas, A., Steinhardt, P. J., and Loeb, A. (2013). Inflationary paradigm in trouble after Planck2013. *arXiv:* 1304.2785.

Ijjas, A., Steinhardt, P. J., and Loeb, A. (2014a). Inflationary schism after Planck2013, *arXiv:* 1402.6980: 1–7.

Isen, A. M. (2001). An Influence of Positive Affect on Decision Making in Complex Situations: Theoretical Issues With Practical Implications. *Journal of Consumer Psychology,* doi:10.1207/S15327663JCP1102_01.

Ivanov, I., Liu, X., Clerkin, S., Schulz, K., Friston, K., Newcorn, J. H., and Fan, J. (2012). Effects of motivation on reward and attentional networks: An fMRI study. *Brain and Behavior, 2* (6): 741–753.

Just A. M., Cherkassky, V. L., Buchweitz, A., Keller, T. A., and Mitchell, T. M. (2014). Identifying Autism from Neural Representations of Social Interactions Neurocognitive Markers of Autism. PLOS, DOI:10.1371/journal. pone. 0113879s 001.

Kaiser, R. I. Stockton, A. M. Kim, Y. S. Jensen, E. C. and Mathies, R. A. (2013) On the formation of dipeptides in Interstellar Model Ices. *Ap. J., 765* (2): 111–120.

Kantak, S. S., Stinear, J. W., Buch, E. R., and Cohen, L. G. (2012). Rewiring the Brain Potential Role of the Premotor Cortex in Motor Control, Learning, and Recovery of Function Following Brain Injury. *Neurorehabilitation and neural repair, 26* (3): 282–292.

Kaufman, M. T., Churchland, M. M., Ryu, S. I., and Shenoy, K. V. (2015). Vacillation, indecision and hesitation in moment-by-moment decoding of monkey motor cortex. *eLife, 4* DOI: 10.7554/eLife.04677

Keizer, K., Lindenberg, S., and Steg, L. (2008). The spreading of disorder. *Science, 322* (5908): 1681-5.

Kim, Y.-H., Yu, R., Kulik, S. P., Shih, Y. H., and Scully, M. O. (1999). A Delayed Choice Quantum Eraser. Phys. Rev. Lett., 84(1), 1–5.

Kline, M. A., and Boyd, R. (2010). Population size predicts technological complexity in Oceania. *Proceedings. Biological Sciences / The Royal Society, 277* (1693): 2559–64.

Kleidon, A. (2010). Life, hierarchy, and the thermodynamic machinery of planet Earth. *Physics of Life Reviews, 7,* 4: 424–460.

Kocsis, S., Braverman, B., Ravets, S., Stevens, M., Mirin, R.,

Shalm, L. and Steinberg, A. (2011). Observing the Average Trajectories of Single Photons in a Two-Slit Interferometer. Science 332, 1170

Koga, Y. (2012). Thermal adaptation of Archeal and bacterial lipid membranes. *Archea,* 10, 1155–61.

Koonin, E. V. (2011) Are There Laws of Genome Evolution? *PLoS Comput Biol., 7* (8),e1002173.

Koonin, E. V. and Wolf, Y. I. (2008). Genomics of bacteria and archaea: the emerging dynamic view of the prokaryotic world. *Nucleic Acids Research, 36* (21): 6688–6719.

Kounios, J., and Beeman, M. (2009). The Aha! Moment: The Cognitive Neuroscience of Insight. *Current Directions in Psychological Science, 18* (4): 210–216.

Kraus, M. W., Huang, C. and Keltner, D. (2010). Tactile Communication, Cooperation, and Performance: An Ethological Study of the NBA. *J. Emotion,* 1–20.

Krioukov, D., Kitsak, M., Sinkovits, R. S., Rideout, D., Meyer, D. and Boguñá, M. (2012). Network Cosmology. *Scientific Reports, 2*: 793–799.

Kwok, S (2009). Organic matter in space: from star dust to the Solar System. *Astrophys Space Sci., 319,* 5–21.

Kwok, S. and Zhang, Y. (2011). Mixed aromatic-aliphatic organic nanoparticles as carriers of unidentified infrared emission features. *Nature (London), 479.* 7371: 80–83.

Laeng B., Waterloo K., Johnsen S. H., Bakke S. J., Låg T., Simonsen S. S., and Høgsæt J. (2007). The eyes remember it: Oculography and pupillometry during recollection in three amnesic patients. *Journal of Cognitive Neuroscience, 19*: 1888–1904.

Leconte, J., Wu, H., Menou, K., and Murray, N. (2015). Asynchronous rotation of Earth-mass planets in the habitable zone of lower-mass stars. *Science,* DOI: 10. 1126/ 1258686.

Leffler, E. M., Bullaughey, K., Matute, D. R., Meyer, W. K., Segurel, L., Venkat, A., and Przeworski, M. (2012). Revisiting an Old Riddle: What Determines Genetic Diversity Levels within Species? *PLoS Biology, 10* (9), e1001388.

Liljenquist, K., Zhong, C.-B., and Galinsky, A. D. (2010). The smell of virtue: clean scents promote reciprocity and charity. Psychological Science, 21(3), 381–383.

Linden, N., Popescu, S., Short, A. J., and Winter, A. (2009). Quantum mechanical evolution towards thermal equilibrium. *Phys. Rev.,* E, 79(6): 061103.

Lobkovsky, A. E., Wolf, Y. I., and Koonin, E. V. (2010). Universal distribution of protein evolution rates as a consequence of protein folding physics. *Proceedings of the National Academy of Sciences of the*

United States of America, 107 (7): 2983–8.

Lupien, SJ, Maheu, F, Tu M, Fiocco, A, and Schramek, TE. (2007). The effects of stress and stress hormones on human cognition. Implications for the field of brain and cognition. *Brain Cogn., 65* (3): 209–237.

Lustenberger, C., Boyle, M. R., Foulser, A. A., Mellin, J. M., and Fröhlich. F., (2015). Functional role of frontal alpha oscillations in creativity. *Cortex,* DOI: 10.1016/j.cortex. 2015.03.012

Makowiecki, K., Harvey, A. R., Sherrard, R. M., and Rodger, J. (2014). Low-Intensity Repetitive Transcranial Magnetic Stimulation Improves Abnormal Visual Cortical Circuit Topography and Upregulates BDNF in Mice. *Journal of Neuroscience, 34* (32): 10780–10792.

Mancini, F., Longo, M. R., Kammers, M. P. M., and Haggard, P. (2011). Visual distortion of body size modulates pain perception. Psychological *Science: APS, 22* (3): 325–330.

Mantini, D., and Vanduffel, W. (2012). Emerging Roles of the Brain's Default mode network. The Neuroscientist.

Martin, J. S., Smith, N., and Francis, C. D. (2013). Removing the entropy from the definition of entropy: clarifying the relationship between evolution, entropy, and the second law of thermodynamics. *Evol. Educ. Outreach, 6* (1): 30.

Martyushev, L. M. (2010). The maximum entropy production principle: two basic questions. *Philosophical Transactions of the Royal Society of London. Series B, Biological Sciences, 365* (1545): 1333–1334.

Mathewson, K., Back, D., Ro, T., Maclin, L. M., Lo, K., and Fabiani, M., (2012). Dynamics of Alpha Control: Preparatory Suppression of Posterior Alpha Oscillations by Frontal Modulators Revealed with Combined EEG and Event-related Optical Signal. *J Neural Cognition,* doi: 10.1162/jocn a 00637.

McCraty, R, and Atkinson, M. (2014). Electrophysiology of intuition: pre-stimulus responses in group and individual participants using a Roulette paradigm. *Global Adv Health Med; 3* (2):16–27.

McNally, L., Brown, S. P., and Jackson, a. L. (2012). Cooperation and the evolution of intelligence. *Proceedings of the Royal Society B: Biological Sciences (April),* doi:10.1098/rspb.2012.0206.

Megidish, E., Halevy, A., Shacham, T., Dvir, T., Dovrat, L., and Eisenberg, H. S. (2013). Entanglement between photons that have never coexisted. *Phys. Rev. Lett., 110 (21):* 210403.

Merlin, F. (2010). Evolutionary Chance Mutation: A Defense Of The Modern Synthesis' Consensus View. *Philos Theor. Biol. 2,* 103.

Moreva, E., Brida, G., Gramegna, M., Giovannetti, V., Maccone, L., and Genovese, M. (2013). Time from quantum entanglement: an experimental illustration, *arXiv,* 1310.4691v1.

Morris, S. C. (2010). Evolution, like any other science is predictable. *Phil. Trans. R. Soc. B., 365,* (1537): 133–145.

Mota, B., and Herculano-Houzel, S., (2015). Cortical folding scales universally with surface area and thickness, not number of neurons. Science, 3 (6243):74-77.

Mrazek MD, Smallwood J, and Schooler JW. (2012). Mindfulness and mind-wandering: finding convergence through opposing constructs. *Emotion, 12* (3): 442–448.

Murakami, M.I., Vicente, G.M., Costa, Z.F. and Mainen (2014). Neural antecedents of self-initiated actions in secondary motor cortex. *Nat Neurosci., 17:* 1574–1582.

Murch, K. (2015). Prediction, Retrodiction, and Smoothing for a Continuously Monitored Supercond. qubit. Bull. Am. Phys. Soc. 60.

Nathaniel V. D. (2011). Thermodynamics and the structure of living systems. Ph.D. thesis, University of Sussex.

Neupert, S. D., and Allaire, J. C. (2012). I think I can, I think I can: Examining the within-person coupling of control beliefs and cognition in older adults. *Psychology and Aging, 2* (2): 145–152.

Nevo, E. and Beiles, A. (2011). Genetic variation in nature. *Scholarpedia, 6*(7): 8821.

Njomboro, P., Deb, S., and Humphreys, G. W. (2012). Apathy and executive functions: insights from brain damage involving the anterior cingulate cortex. BMJ case reports, bcr0220125934. doi: 10.1136/bcr 5934.

Oei, N. Y. L., Rombouts, S. A.R.B., Soeter, R. P., van Gerven, J. M and Both, S. (2012). Dopamine Modulates Reward System Activity During Subconscious Processing of Sexual Stimuli. *Neuropsychopharmacology, 37:* 1729–1737.

Oohashi, T., Nishina, E., Honda, M., Yonekura, Y., Fuwamoto, Y., Kawai, N., and Shibasaki, H. (2000). Inaudible high-frequency sounds affect brain activity: hypersonic effect. *Journal of neurophysiology, 83:* 3548–3558.

Ohman A., and Mineka S. (2001). Fears, phobias, and preparedness: toward an evolved module of fear and fear learning. *Psychol Rev., 108* (3), 483–522.

Overmier, J. B. and Seligman, M.E.P. (1967). Effects of inescapable shock upon subsequent escape and avoidance responding.
Journal of Comparative and Physiological Psychology, 63: 28–33.

Öberg, K. I., Guzmán, V. V., Furuya, K., Qi, C., Aikawa, Y., Andrews, S. M., Loomis, R., and Wilner. D. J., (2015). The cometary composition of a protoplanetary disk as revealed by complex cyanides. *Nature, 520,* 198–201.

Padmanabhan, T. (2011). Gravity as an emergent phenomenon: Conceptual aspects. AIP *Conf. Proc., 1458:* 238–252.

Padmanabhan, T. (2010). Thermodynamical Aspects of Gravity: New insights. *Rep. Prog. Phys.,* 73,046901.

Padmala, S., and Pessoa, L. (2011). Reward Reduces Conflict by Enhancing Attentional Control and Biasing Visual Cortical Processing. *Journal of Cognitive Neuroscience,* doi:10.1162/jocn_a_00011.

Penttonen, M. and Buzsaki, Gy. (2003). Natural logarithmic relationship between brain oscillators. *Thalamus & Related Systems, 2* (2): 145–152.

Peplow, M. (2014). Peer review — reviewed. Nat. News. doi:10.1038/ nature. 16629

Pessoa, L. (2005). To what extent are emotional visual stimuli processed without attention and awareness, Current Opinion in Neurobiology, 15: 188–196.

Piketty, T. (2014). "Capital in a Twenty-First Century" Belknap Press.

Pizzarello, S (2006). The Chemistry of Life's Origin: A Carbonaceous Meteorite Perspective. *Acc. Chem. Res., 39* (4): 231–237.

Pizzarello, S., and Shock, E. (2010). The organic composition of carbonaceous meteorites: the evolutionary story ahead of biochemistry. *Cold Spring Harbor perspectives in biology,* doi:10.1101/.

Pothos, E. M. and Busemeyer, J. R. (2009). A quantum probability model explanation for violations of 'l' derationacision theory. *Proceedings of the Royal Society, B, 276* (1665): 2171–2178.

Pothos, E. M. and Busemeyer, J. R. (2013). Can quantum probability provide a new direction for cognitive modeling? *Behavioral and Brain Sciences, 36* (03): 255–274.

Powell, R. (2011). The Future of Human Evolution. *The British Journal for the Philosophy of Science, 63*(1): 145–175.

Pulcu, E., Lythe, K., Elliott, R., Green, S., Moll, J., Deakin, J. F. W., and Zahn, R. (2014a). Increased amygdala response to shame in remitted major depressive disorder. *PLoS ONE, 9* (1).

Pulcu, E., Zahn, R., Moll, J., Trotter, P. D., Thomas, E. J., Juhasz, G., ... Elliott, R. (2014b). Enhanced subgenual cingulate response to altruistic decisions in remitted major depressive disorder. *NeuroImage: Clinical,* 4: 701–710.

Quian Quiroga, R., Kraskov, A., Mormann, F., Fried, I., and Koch, C. (2014). Single-Cell Responses to Face Adaptation in the Human Medial Temporal Lobe. *Neuron, 84,* 1–7.

Radman, T., Su, Y., An, J. H., Parra, L. C., and Bikson, M. (2007). Spike timing amplifies the effect of electric fields on neurons: implications for endogenous field effects. *The Journal of Neuroscience, 27* (11): 3030–3036.

Raichle, M. E., and Snyder, A. Z. (2007). A default mode of brain function: A brief history of an evolving idea. *NeuroImage, 37* (4): 1083–1090.

Ramos, Jo Ao, and Torgler, B. (2010). Are Academics Messy? Testing the Broken Windows Theory with a Field Experiment in the Work Environment. *Technology, 8* (3): 563–577.

Rand, D., Green, J., and Novak, M. (2012). Spontaneous giving and calculated greed. *Nature, 489*: 427–430.

Rudd, M., Aaker J., and Vohs. K. (2012). Awe Expands People's Perception of Time, Alters Decision Making, and Enhances Well-Being. *Psychological Science, 23* (10): 1130-6.

Sadaghiani, S., Hesselmann, G., and Kleinschmidt, A. (2009). Distributed and antagonistic contributions of ongoing activity fluctuations to auditory stimulus detection. *The Journal of Neuroscience, 29* (42): 13410–13417.

Saphiro J. A. (2011). Evolution: A view from the twentieth century. FT Press.

Savage, L. J. (1954). The Foundations of Statistics: John Wiley and Sons.

Seligman, M.E.P. and Maier, S.F. (1967). "Failure to escape traumatic shock." *Journal of Experimental Psychology, 74*: 1–9.

Sellaro, R., Derks, B., Nitsche, M. A., Hommel, B., Wildenberg, W. P.M. van den, Dam, K. van, and Colzato. L. S., (2015). Reducing prejudice through brain stimulation. *Brain Stimulation*, DOI: 10.1016/j. brs.2015.04.003

Seo, D., Patrick, C. J., and Kennealy, P. J. (2008). Role of Serotonin and Dopamine System Interactions in the Neurobiology of Impuslive Aggression and its Comorbidity with other Clinical Disorders. *Aggression and Violent Behavior, 13* (5): 383–395.

Schooler, J. W., Smallwood, J., Christoff, K., Handy, T. C., Reichle, E. D., and Sayette, M. A. (2011). Meta-awareness, perceptual decoupling and the wandering mind. *Trends in Cognitive Sciences, 15* (7): 319–326.

Schrödinger, E (1945). What Is Life? The Physical Aspect of the Living Cell. The University Press.

Schultz, W. (2007). Reward signals. Scholarpedia, 2(6): 2184.

Scully, M. O. and Druhl, K. (1982). Quantum eraser – A proposed photon correlation experiment concerning observation and 'delayed choice' in quantum mechanics. *Phys. Rev. A, 25*: 2208–2213

Simon-Thomas, E. R., Godzik, J., Castle, E., Antonenko, O., Ponz, A., Kogan, A., and Keltner, D. J. (2011). An fMRI study of caring vs self-focus during induced compassion and pride. *Social Cognitive and Affective Neuroscience (2)*, 212.

Smallwood, J., Fitzgerald, A., Miles, Lynden K, and Phillips, L. H. (2009). Shifting moods, wandering minds: negative moods lead the mind to wander. *Emotion, 9* (2): 271–276.

Soon, C. S., Brass, M., Heinze, H. J. and Haynes, J. D. (2008). Unconscious

Determinants of Free Decisions in the Human Brain. *Nature Neuroscience*, *11*, 5: 543–545.

Soon C. S., He A. H., Bode S., and Haynes J. D. (2013). Predicting free choices for abstract intentions. *Proc Natl Acad Sci, 110* (15): 6217–6222.

Stellar, J. E.; John-Henderson, N; Anderson, C. L.; Gordon, A. M.; McNeil, G. D.; and Keltner, D. (2015). Positive Affect and Markers of Inflammation: Discrete Positive Emotions Predict Lower Levels of Inflammatory Cytokines. *Emotion*, (2): 129-33.

Steptoe, A. and Wardle, J. (2005). Positive affect and biological function in everyday life. *Neurobiol Aging, 26,* Suppl. 1:108–12.

Storm, B. C., and Stone. S. M. (2014). Saving-Enhanced Memory. The Benefits of Saving on the Learning and Remembering of New Information. *Psychological Science*, DOI: 10.1177/0956797614559285.

Strominger, A., Yau, Shing-Tung and Zaslow, E. (1996). Mirror symmetry is T-duality. *Nuclear Phys., B, 479,* (1-2): 243–259.

Stueckelberg, E. (1941). *Helv. Phis. Acta, 14*: 51–80.

Szendro, I. G., Franke, J., de Visser, J. A. G. M., and Krug, J. (2013). Predictability of evolution depends nonmonotonically on population size. *Proceedings of the National Academy of Sciences of the United States of America, 110* (2): 571–6.

Susskind, L. (1994). The World as a Hologram. *arXiv:* hep-th/9409089v2. 1–34.

Tang, A., Jackson, D., Hobbs, J., Chen, W., Smith, J. L., Patel, H., Prieto, A., Petrusca, D., Grivich, M. I., Sher, A., Hottowy, P., Dabrowski, W., Litke, A. M., and Beggs, J. M. (2008). A maximum entropy model applied to spatial and temporal correlations from cortical networks in vitro. *J. Neurosci., 28*: 505–518.

Toups, M. A., Kitchen, A., Light, J. E., and Reed, D. L. (2011). Origin of clothing lice indicates early clothing use by anatomically modern humans in Africa. *Molecular Biology and Evolution, 28* (1): 29-32.

Toyabe, S., Sagawa, T., Ueda, M., Muneyuki, E. and Sano, M. (2010). Experimental demonstration of information-to-energy conversion and validation of the generalized Jarzynski equality. *Nature Physics*, doi: 10.1038/NPHYS1821.

Treadway, M. T., Buckholtz, J. W., Cowan, R. L., Woodward, N. D., Li, R., Ansari, M. S., ... Zald, D. H. (2012). Dopaminergic mechanisms of individual differences in human effort-based decision-making. *The Journal of Neuroscience, 32*, 18: 6170–6176.

Tversky A., and Shafir, E. (1992). The disjunction effect in choice under uncertainty. *Psychological Science, 3*: 305–309.

Unruh, W. G. (1992). Thermal bath and decoherence of Rindler spacetimes. *Phys. Rev.,* D46: 3271.

Urry, H. L., and Gross, J. J. (2010). Emotion Regulation in Older Age. *Current Directions in Psychological Science, 19* (6): 352–357.

Van Dillen, L. F. van der Wal, R. C., and van den Bos,K. (2012). On the Role of Attention and Emotion in Morality: Attentional Control Modulates Unrelated Disgust in Moral Judgments. *Pers Soc Psychol Bull, 38*: 1222–1231.

Van Gaal, S., De Lange, F. P., and Cohen, M. X. (2012). The role of consciousness in cognitive control and decision making. *Frontiers in Human Neuroscience, 6*: 1–15.

Veneziano, G. (1968). Construction of a crossing—symmetric, Regge behaved amplitude for linearly rising trajectories. *Nuovo Cimento* A, 57: 190–197.

Verlinde, E. (2010). On the Origin of Gravity and the Laws of Newton. *J. High Energy Phys.,* 1104: 029.

Vetere, G., Restivo, L., Cole, C. J., Ross, P. J., Ammassari-Teule, M., Josselyn, S. A., and Frankland, P. W. (2011). Spine growth in the anterior cingulate cortex is necessary for the consolidation of contextual fear memory. *Proceedings of the National Academy of Sciences, 108* (20): 8456–8460.

Vishnoi, A., Kryazhimskiy, S., Bazykin, G. A., Hannenhalli, S., and Plotkin, J. B. (2010). Young proteins experience more variable selection pressures than old proteins. *Genome Research, 20* (11): 1574–1581.

Wang, G. M., Sevick, E. M., Mittag, E., Searles, D. J., and Evans, D. J. (2002). Experimental demonstration of violations of the second law of thermodynamics for small systems and short time scales. *Phys. Rev. Lett., 89* (5): 050601.

Warnefors, M., and Eyre-Walker, A. (2011). The accumulation of gene regulation through time. *Genome Biology and Evolution, 3*, 667–673.

Weiss, S. A., and Faber, D. S. (2010). Field effects in the CNS play functional roles. *Frontiers in Neural Circuits, 4*: 15–30.

Weiskrantz, L. (1990). Outlooks for blindsight: Explicit methodologies for implicit processes. The Ferrier Lecture. *Proceedings of the Royal Society Series B. London, 239*: 247–278.

Weiskrantz, L. (1998). Pupillary responses with and without awareness in blindsight. *Consciousness and Cognition, 7*: 324–326.

Weygaert, R. (2007). Voronoi Tessellations and the Cosmic Web: Spatial Patterns and Clustering across the Universe The cosmic web: geometric analysis. *Astrophys., 1*: 070708.1441.

Wissner-Gross, A. D., and Freer, C. E. (2013). Causal Entropic Forces. *Physical Review Letters, 110*(16): 168702.

Widrow, L. M. (2002). Origin of Galactic and Extragalactic Magnetic Fields, *Rev. Mod. Phys., 74:* 775–823.

Wolfram, S. (2002). A new kind of science. (US) Wolfram Media. Inc.

Wyart, V., and Sergent, C. (2009). The phase of ongoing EEG oscillations uncovers the fine temporal structure of conscious perception. *The Journal of neuroscience, 29* (41): 12839–12841.

Yamada, Y., and Kawabe, T. (2011). Emotion colors time perception unconsciously. Consciousness and Cognition, Elsevier Inc. 20(4), 1–7.

Xu, S. Z. Yan, K. Jang, W. Huang... Rogers, J.A. (2015). Assembly of micro/nanomaterials into complex, three-dimensional architectures by compressive buckling. *Science, 347* (6218): 154–159.

Yoshimura, S., Okamoto, Y., Onoda, K., Matsunaga, M., Ueda, K., Suzuki, S., and Yamawaki, S. (2010). Rostral anterior cingulate cortex activity mediates the relationship between the depressive symptoms and the medial prefrontal cortex activity, *Journal of Affective Disorders,122,* 1–2: 76–85.

Zeeman, E.C. (1976). Catastrophe Theory, *Scientific American, 234* (4): 65–70, 75–83.

Glossary

Acceleration Acceleration is the rate of change of an object's velocity over time. Like gravity, acceleration slows time, although for contrary reasons. Acceleration takes one toward a SF with lesser curvature (i.e., a „younger" field), reducing the frequency of decoherence and reversing time's flow. The combined effect slows the inner clock of material systems.

Antimatter The Feynman–Stueckelberg interpretation states that antimatter and antiparticles are regular particles traveling backward in time due to up spin decoherence.

Black hole Regions of great entropy, typified by an asymptotic inhibition of decoherence. The quantum frequencies approach infinity, which cancels their energy differences and reduces them into their information saturated and impenetrable event horizon.

Calabi–Yau space (Calabi–Yau torus or simply „torus") The Calabi–Yau space defines a microdimensional space, which was formed in the Big Bang. Insulated from the SF by its manifold (horizon), the torus forms standing waves (i.e., quantum energy) without spatial limitations. During interaction, the ratio of manifold energy to quantum energy changes inversely, but for outside observers the particle remains constant. The increasing frequencies accumulate information (information memory or positive time), which weakens the manifold. In the opposite case, an increase in manifold energy constitutes negative time. The energy imbalance of MiDV is transmitted by the field.

Charge Charge is the uninsulated intrinsic angular momentum of the originating particle.

Cosmic void A cosmic void is formed by the negative spatial curvature of the manifold-energy-rich Calabi–Yau space. The negative spatial curvature expels matter, thus functions as anti-gravity. Cosmic voids bulge into the fourth dimension by expansion.

Dark energy Dark energy is supposed to account for the accelerating expansion of the universe. However, the existence of white holes naturally opens an expansionary energy in the cosmos and presents an operational possibility for the universe without requiring dark energy.

Dark matter Dark matter was introduced to explain the mass discrepancies experienced on large scales in the universe. Nevertheless, it can be shown that the expansion of the universe generates pressure within gravitational regions, leading to the appearance of extra matter.

Decoherence (interaction) Interaction changes the standing-wave frequencies of the microdimensions, and the corresponding manifold energy by Lorentz transformation.

Decoherence (Interaction between temporal fermions) The emotional mind's interaction with the energy of time changes mental volume in discrete increments. It produces mental expansion (mental energy) or contraction (information memory) according to the Heisenberg uncertainty principle.

Default mode network (DMN) The low-frequency (0.1 Hz) oscillations of specific brain regions that form the brain's energy-neutral state.

Ego Insulated quantum energy forms an energy unit called the self, or ego. Through sensory processing it identifies itself with the body and becomes the source of self-awareness. The ego's aim to protect the self from immediate pain, gives rise to a self-centered orientation manifested as temporal weight.

Emotional charge An intrinsic mental orientation between past and future that fuels an opinion (attitude) for almost every question.

Emotional electron Animals with well-formed cerebrums form energy-neutral standing waves and can be considered emotional electrons. Temporal energy imbalance (energy vacuum) is felt as emotion. The manifold energy (i.e., subconscious) allows learning through the accumulation of experience.

Emotional distance A conceptual (temporal) distance between emotional beings, forming a gut feeling of comfort. Any deviation from this distance would trigger emotion. Violation of the emotional (temporal) distance triggers the Pauli exclusion principle, which is the need for separation.

Emotional gravity (Temporal gravity) Positive TF curvature forms emotional gravity, which constricts time and enhances the willingness for interaction. Species, individuals and even companies over time inherently gravitate toward a TF curvature that is congruent with their temporal weight.

Emotional mass see **temporal mass**

Emotional temperature Emotional temperature measures the extent of negativity, which evolves dependent on situations or relationships.

Energy-neutral unit Elementary fermions maintain energy neutrality over space or time analogue to the mechanism of „static" time. The universe gives rise to submanifolds with similar operation and structures. Thus ecosystems and societies preserve energy neutrality.

Energy vacuum Energy vacuum is a spatial, or temporal energy imbalance of the Calabi–Yau space.   Energy vacuum is abhorred; therefore it is eliminated during subsequent interaction. MiDV (MiDT) forms within the Calabi–Yau torus microdimensions, whereas MaDV (MaDT) occupies the macrodimensions, which are outside the torus. During interaction, reciprocal modification of MiDV (MiDT) and MaDV (MaDT) ensures conservation of energy.

Entanglement see **quantum entanglement**

Entropy Entropy is the stability of non interaction, due to the strength of the connection between the quantum wave and the manifold. A strong connection corresponds to high entropy, whereas within low entropy conditions a weak connection ensures the willingness for interaction. High-entropy regions are the poles of the universe. Thus, energetic movement toward either of the poles increases entropy.

Entropic force   The entropy-maximizing drive of positive spatial curvature leads to the second law of thermodynamics.

Entropic pressure The source of evolutionary change; proportional to the entropy, the dynamism (physical, and chemical conditions) of the gravitational environment and inversely proportional to the relative concentration of membranous organisms. The increasing entropy of the environment is balanced by the increasing order inside the membrane.

Evolutionary weak nuclear force (temporal weak nuclear force) The environment's ability to renew itself after massive degradation occurs by the inversion of the temporal curvature by the evolutionary weak nuclear force. The entropic drive of the degraded environment fuels evolutionary renewal. For example genetic changes give rise to new functions and the waste of the previous era will nourish and support the regenerating ecosystem of a new evolutionary period.

Fermion (material fermion)   The interacting temporal Calabi–Yau space gives rise to matter. See particle.

Genetic diversity Genetic diversity is proportional to the mental entropy of the population. Artificially bred populations are exceptions.

Gravity Gravity is formed by positive SF curvature. The pressure of the SF strength is mistakenly interpreted as the pull of gravity.

Gravitational potential see **quantum energy**

Gravity gradient tensor (second derivative of gravitational potential) see **spatial field**

Heisenberg uncertainly principle The insulated Calabi-Yau torus determines discrete frequencies and discrete energy changes during interaction. So the frequency shift of the particle can only occur into either one of two energy state possibilities (probability amplitudes), which is expressed in the Heisenberg uncertainty principle.

Information memory or **Information** Accumulation of information (down spin decoherence) weakens the manifold of fermions by eliminating the degrees of freedom.

Manifold energy (energy potential) The universe is divided into micro- and macrodimensions by the horizon (manifold). The manifold represent the possible energy states of the system: it is a dynamic variable, which can transform into information and back through interaction. Manifold energy can be considered a potential energy, which inhibits the singularity of wave-function collapse. Maniford energy is equivalent to „quantum potential" in Bohmian mechanics.

Manifold energy (in the mind) The mental manifold (subconscious) is the horizon that separates the macro- and the microdimensions. The cortex records quantum activity as the holographic (emotional) history of the organism, represented by neuronal organization. The degrees of mental freedom are the extent of mental openness (manifold energy), freshness, and trust. The open mind sees opportunities, whereas the temporally constricted mind sees limitations.

Macrodimensional vacuum or **macrodimensional time** (**MaDV** or **MaDT**) The spatial (or temporal vacuum) of up spin decoherence insulates the torus from the field. Since it cannot connect to the SF or TF, it does not get transmitted by them. As a consequence, the MaDV (MaDT) is eliminated instantaneously, restoring the energy neutrality of the torus.

Matter Manifested by the interacting Calabi–Yau torus, it is the source of sensory experience.

Mass Mass has a geometric nature. More elongated quantum waves form greater connection to the field, corresponding to greater mass.

Mental energy (Spearman) The source of intellectual abilities; corresponds to degrees of freedom. By forming all inherent, automatic capacities of the organism, mental energy is mental flexibility, congruence, clear conscience, forming trust and confidence.

Mental expansion Mental expansion increases the degrees of freedom through mental connections, new vision, creativity, forming a liberating, joyful feeling.

Mental entropy Mental entropy expresses probability of mental change. High mental entropy is characterized by lack of change. High manifold energy torus does not have enough quantum energy for interaction, whereas great quantum energy is characterized by frequent back-and-forth transformations (the Weyl tensor), which have no net effect. In addition, the energy exchange is too small to exert change. Interaction in the low-entropy mind occurs according to the Pauli exclusion principle.

Mental entropy (in evolution)   Low mental entropy is formed within a Euclidean TF, whereas a curving TF determines high mental entropy. As a result, high mental entropy can be either mental openness or trust (high mental energy within negative field curvature), which allows permissive mating in animals and leads to great genetic diversity, or it can be an agitated search for survival (immense positive temporal curvature induces high quantum energy). Low mental entropy is characterized by luck of trust. The great selection pressure through complex courting rituals keeps the genetic diversity low.

Mental contraction   Mental contraction is the decrease of mental energy (degrees of freedom) through down spin decoherence. It leads to uncertainty, which curtails the possibilities available for the organism. It increases information content, which overwhelms with unnecessary details, so the focus narrows, leading to impatience and failures.

Microdimensional vacuum or **microdimensional time (MiDV or MiDT)**   Down spin particles form an energy vacuum inside the Calabi–Yau torus. MiDV and MiDT are transmitted by the SF or the TF, respectively, as wave packets of photons (or temporal photons). The photon frequency reflects the energy difference of standing waves.

Orthogonal   Just like electricity and electromagnetism, which are related by an orthogonal transformation expressed by multiplication by the imaginary number i, interacting fields of the universe are always oriented orthogonally to each other. The number of orthogonal fields continually doubles during evolution. Some examples are space versus time, quantum versus manifold energies, organization of proteins versus genetic complexity, bodily organization versus neural complexity, biosphere, social, or economic organization versus temporal gravity.

Particle (elementary particle)   Used interchangeably with Calabi–Yau torus in the text. Only an interacting Calabi–Yau torus can be experienced by the sensory organs and measuring equipment. Interacting torus is commonly called matter. See also Calabi–Yau space.

Pauli exclusion principle   The principle that recognizes that the minimal-energy formation between neighboring particles corresponds to opposite, up- and down spin configurations: entanglement.

Photon   The spatial contraction of a down spin particle is transmitted across the SF (TF) as a wave packet (spatial or temporal boson).

Population entropy see **societal entropy**

Quantum energy The formation of the microdimensions turns particle vibrations into standing waves. Changes of standing-wave frequencies would free or absorb energy in quanta. During down spin decoherence the quantum energy forms higher-frequency standing wave overtones, which increase the particle's willingness for interaction. Up spin is the reverse process.

Quantum entanglement Entanglement is a common wave function of complementary sister particles which appear to be independent. The energy distribution can be manipulated between the entangled particles until the next decoherence (examination), which is the next time moment and irreversibly separates the common wave function.

Shame Falling in physical space occurs due to gravity. Similarly, shame can be understood as an insecure, painful feeling of a sudden fall of confidence due a shift of temporal gravity.

Societal entropy (Population entropy) Societal entropy is a measure of societal stability. High social entropy ensures great stability, whereas low-entropy conditions are wrought by constant change. In newly formed societies, high quantum energy corresponds to chaotic circumstances, insecurity, and fear. The downtrodden population is powerless to instigate change. In low-entropy conditions gravitational interactions occur according to the Pauli exclusion principle. Mature, secure societies form highly entropy (manifold energy) through civil confidence, and trust in social and political institutions, which eliminates the desire for change.

Subconscious see **mental manifold**

Spatial field (SF) The SF is made up of expansion and contraction (compactification) forces that change inversely to each other. Expansionary energy decreases field strength and forms the negative volume of white holes by convention. Contraction eliminates volume by increasing field curvature (field strength). By convention, contraction agrees with the positive space of the rolled-up dimensions. Within galactic regions, the equal balance of expansion and contraction energies forms Euclidean field. The field curvature corresponds to the gravity gradient tensor.

Spatial Calabi–Yau space see **emotional (temporal) fermion**

Spin Spin is the intrinsic angular momentum of fermions. Decreasing frequencies contract the torus and form up spin, whereas the increasing frequencies of down spin expand the horizon of the torus, thus fermions are half-spin particles. The orthogonality of the quantum wave and the SF turns spin into a vector quantity.

Static time Page and Wootters showed how entanglement can be used to measure time in 1983. Time is an emergent phenomenon of a clock like evolution of entangled particles (i.e., inside observers). For outside observers the universe remains static and timeless, just as the Wheeler-DeWitt equations predict. The simple and powerful idea was proven in 2013 by Moreva and colleagues.

Stress Excessive temporal curvature enhances brain frequencies and creates temporal confinement, impatience, which leads to the sense of lack of time.

Temporal fermion (temporal elementary particle) The separation of the SF within microdimensions forms a symmetric energy structure to matter: the temporal elementary particle.

Temporal field (TF) The existence of the Calabi–Yau torus turns time discrete and time reversal is a function of interaction. The manifold information content corresponds to the age of its local coordinate. By convention, negative temporal energy points toward white holes (time zero); positive temporal energy is typical of black holes (time is infinite).

Temporal field (in the mind and evolution) The chain of events and their responses by the organism or species forms the temporal field, which becomes the single most important determinant, if not the only determinant of behavior. We cannot ignore our emotional history: it directs our lives impeccably and deterministically.

Temporal gravity Temporal gravity is the positive curvature of the TF; it corresponds to the emotional hold of the individual by family, things, society, culture, and historical circumstance.

Temporal gravity (in evolution) Temporal mass is typical of the species and it determines its place within the TF curvature layers of the ecosystem. High temporal gravity is taxing and makes survival difficult. Thus, even the smallest changes in temperature or other ecological parameters can lead to extinction.

Temporal mass or **emotional mass** (of a species) Just like the mass of matter fermions, temporal mass is proportional to the elongation of the quantum energy. The temporal gravity and the temporal mass must be congruent for stability; therefore, the survival of a species is limited to an evolutionary period with compatible temporal gravity. Emotional organisms cannot have zero temporal mass.

Temporal weak nuclear force see **evolutionary weak nuclear force**

Temporal weight (emotional weight) The temporal mass is constant and characteristic of the species, but the temporal weight is typical of the individual or group and forms over time interdependent with the environment. Hence, temporal weight is individual temporal curvature, which is inversely proportional to mental energy. The ego's self-centered orientation leads to its increasing reliance on the environment, which increases the temporal weight. People with great emotional weight are highly conservative and form close, even suffocating personal relationships. Their extreme emotional dependence on others can lead to abusive relationships (where they can be either the abuser or the abused). Thus, the termination of the relationship (the removal of the emotional support) collapses the TF and can lead to depression. Small temporal weight corresponds to openness, confidence, which permits emotional independence and freedom.

Time The information content of the local manifold. The change of energy in one decoherence determines the speed of inner clocks. Material interaction with the SF consumes space to produce time: manifold information accumulation. Thus, time (information) and space (manifold energy) are convertible energies.

Temporal energy see **temporal field**

Weight Each particle forms a field curvature in proportion to its own mass. However, interaction of the particle and the SF forms weight. Weight is the force that restores the field curvature, free of the particle.

Weyl tensor The immense positive field curvature (elliptic geometry) close to black holes squeezes space, so quantum energy materializes as a shear stress.

Weyl tensor (In temporal dimension) The positive TF curvature constricts mental focus, overloads the mind with details, leading to the feeling of lack of time, called stress. It manifests itself as lack of patience, a tendency for aggravation, contradictions, and a tendency to criticize (shear stress).

White hole Areas with pronounced negative field curvature inhibits interaction. The information-free Calabi–Yau torus is contracted, encouraging spatial expansion, which is known as the Ricci scalar.

Index

Locator numbers for Notes (either in boxes or at the end of Chapters One and Two) is followed by 'n'.

About the Author

Eva Deli is a Hungarian American with a background in cell biology and the visual arts. In cancer research she has coauthored fifteen peer reviewed publications and as an artist she had an active exhibition schedule for over a decade. Her creative energy has been shaped by a powerful visual comprehension combined with scientific rigor. She is a self-taught scholar of theoretical physics, neurology and evolutionary biology. Eva is also a keen observer of people. Her idea about the nature of consciousness is part of a cohesive, encompassing hypothesis on the natural world and evolution. She can be contacted at www.evadeli.com and www.thescienceofconsiousness.com.

Made in the USA
Middletown, DE
14 May 2021

39666812R00124